teach® yourself

geology

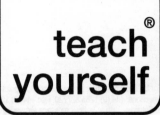

geology
david a. rothery

For UK order enquiries: please contact Bookpoint Ltd, 130 Milton Park, Abingdon, Oxon OX14 4SB. Telephone: +44 (0) 1235 827720. Fax: +44 (0) 1235 400454. Lines are open 09.00–17.00, Monday to Saturday, with a 24-hour message answering service. Details about our titles and how to order are available at www.teachyourself.co.uk

For USA order enquiries: please contact McGraw-Hill Customer Services, PO Box 545, Blacklick, OH 43004-0545, USA. Telephone: 1-800-722-4726. Fax: 1-614-755-5645.

For Canada order enquiries: please contact McGraw-Hill Ryerson Ltd, 300 Water St, Whitby, Ontario L1N 9B6, Canada. Telephone: 905 430 5000. Fax: 905 430 5020.

Long renowned as the authoritative source for self-guided learning – with more than 50 million copies sold worldwide – the **teach yourself** series includes over 500 titles in the fields of languages, crafts, hobbies, business, computing and education.

British Library Cataloguing in Publication Data: a catalogue record for this title is available from the British Library.

Library of Congress Catalog Card Number: on file.

First published in UK 1997 by Hodder Education, part of Hachette Livre UK, 338 Euston Road, London, NW1 3BH.

First published in US 1997 by The McGraw-Hill Companies, Inc.

This edition published 2008.

The **teach yourself** name is a registered trade mark of Hodder Headline.

Copyright © 1997, 2003, 2008 David Rothery

Typeset by Transet Limited, Coventry.
Printed in Great Britain for Hodder Education, an Hachette Livre UK Company, 338 Euston Road, London NW1 3BH, by Cox & Wyman Ltd, Reading, Berkshire.

The publisher has used its best endeavours to ensure that the URLs for external websites referred to in this book are correct and active at the time of going to press. However, the publisher and the author have no responsibility for the websites and can make no guarantee that a site will remain live or that the content will remain relevant, decent or appropriate.

Hachette Livre UK's policy is to use papers that are natural, renewable and recyclable products and made from wood grown in sustainable forests. The logging and manufacturing processes are expected to conform to the environmental regulations of the country of origin.

Impression number   10 9 8 7 6 5 4 3 2
Year   2012 2011 2010 2009 2008

contents

acknowledgements

I am grateful to Jayne Walker for her prompt and sound advice on the draft of the first edition, and to friends and colleagues who so willingly took the time to discuss many specific points with me while the book was in preparation, and to my family for putting up with my cries of 'go away' when they attempted to interrupt me. This new edition has been substantially revised, mostly in small matters except for the Solar System geology in Chapter 14 and fuller mention of human-induced global warming when discussing the geological context of climate change. The book now has an expanded glossary. When referring to sufficiently long periods of time or very ancient dates, I write 'billions of years' rather than 'thousands of millions of years'.

The diagrams in this book have been specially prepared, and most of the photographs are my own. With those exceptions, the illustrations are from the following sources:

Figure 3.3 United States Geological Survey; Figure 5.6 Stuart Hall; Figure 5.12 Peter Francis; Figure 12.5 Phil Allen (Production Geoscience Ltd) and Simon Stewart (BP); Figures 13.3–13.6 are modified from various sources, but draw heavily on the work of C. R. Scotese and co-workers; Figures 14.2–14.4, 14.8–14.11, 14.13, 14.15, 14.16 NASA, Figure 14.5 Don P. Mitchell and Yuri Getkin, Figure 14.7 ESA/DLR/FU Berlin (G Neukum).

Plate 1 JAMSTEC/Earth Simulator Center; Plate 13 John Watson (copyright Open University); Plate 14 NASA/JPL/Cornell, Plate 15 NASA.

01

introduction: a mighty matter of legend

In this chapter you will learn:

- about the scope of this book, and how most chapters are intended to help you understand particular processes
- the three main types of rock, and how they are related by the 'rock cycle'.

'The green earth, say you? That is a mighty matter of legend, though you tread it under the light of day!' Although taken from a work of fiction*, about a fictional situation, those words may be taken as the theme for this book. The Earth is a mighty matter of legend. Its origin and the events that have taken place in its history do indeed form a grand and epic tale. Moreover, various parts of this story have been interpreted in many ways; some of these are contradictory, and others are now thought to be untrue.

I should also point out that some of what Earth scientists now believe fairly firmly will undoubtedly turn out to be wrong, or perhaps only partly correct. However, I think it is fair to claim that we understand very well (for the purpose of an introductory book such as this) most of the processes that shape the materials at and immediately beneath the Earth's surface. Probably, what we lack most is a fully developed understanding of the interdependence of all these processes. For example, over geological timescales measured in tens of thousands or millions of years the Earth's climate has changed many times. It is changing today, and the consensus among scientists is that humans are to blame for the current rate of global warming because of the gases we have released into the atmosphere – chiefly carbon dioxide – through burning coal and oil and various industrial processes. However, other factors drive climate change too: changes in the tilt of the Earth's axis and slight variations in the shape of its orbit, minor fluctuations in the energy output of the Sun, the slow drift of the continents that forces readjustment of the ocean circulation pattern, the rise and fall of mountain belts, catastrophes such as giant volcanic eruptions and impacts by asteroids and comets, and even the emergence and evolution of life itself have all played some role in driving the temperature and climate of the Earth in one direction or another. Just how significant human-induced global warming will be in the grand scheme of things remains to be seen, though I think it is likely to impact severely on the way of life in many parts of the globe.

Climate, in turn, influences how fast and in what manner surface rock is worn away and redeposited elsewhere as sediment, so all the factors listed above must leave some kind of imprint in the history of our planet as revealed by a study of the rocks. The complexities are boundless, but nevertheless I believe we now have a fairly complete understanding of how the Earth as a whole functions, at least in broad outline.

* Aragorn to Eomer, *The Two Towers*, J. R. R. Tolkien, 1954

People may become interested in geology, which is the scientific study of the Earth, for a variety of reasons. Many are intrigued by fossils, or pebbles that they pick up on the beach, and wonder what they mean. Others want to know more about earthquakes and volcanoes reported in the news, and whether such events are likely in their own neighbourhood. Some may be curious about where the oil comes from that powers their car, or what their house bricks are made of. And some are just overawed by the grandeur of the mountains or impressed by the pictures of the Earth and of other Earth-like bodies sent back by spaceprobes, and decide they want to find out more.

Whatever your own reason for opening this book, I hope you will find what you want. I have tried to address all these concerns, and others. Most chapters aim to help you understand a particular type of geological process. You don't have to read these chapters in the sequence in which they are printed, but it will be easier if you do. This is because I have attempted to build up the use of terminology (and geology is riddled with strange sounding words!) in a coherent fashion. Notable exceptions to the process-orientated approach are the next chapter, which is a descriptive account of the composition of the Earth, and the final chapter, which deals with fieldwork and the simple equipment you will find useful if you decide to go and look at rocks and fossils for yourself. There is a glossary at the end, where the most important terms (introduced in **bold** in the text) are defined.

Rock types and the rock cycle

It is helpful at this stage to define terms that describe each of the three main kinds of rock. **Igneous** rock is material that was once molten; it usually contains **crystals** that grew within this molten material as it cooled. **Metamorphic** rock is made when heat or pressure (or both) causes a pre-existing rock to recrystallize, but without melting. **Sedimentary** rock forms at the surface, by deposition of detrital grains or by precipitation from solution in water.

A simplified view of the origins of these rock types is shown in Figure 1.1, which illustrates what is known as the **rock cycle**. This includes the erosion, transport, deposition, burial, heating, deformation, melting, cooling and exhumation of rock or rock fragments. Rock on high ground is worn to fragments by frost, rain and wind, and transported downhill to a place where these fragments are deposited and buried. This is the erosional and depositional part of the rock cycle. If these processes were the sole agent of change, the ultimate fate of the Earth's surface would be for hills and mountains to be worn away, until eventually everywhere was flat and buried by mud. However, the Earth is a dynamic planet and forces of deformation and uplift are at work that continually bring igneous and metamorphic rock, and hardened sediment, to the surface. This provides an unending supply of material to be worked on by the forces of erosion, and so the rock cycle continues.

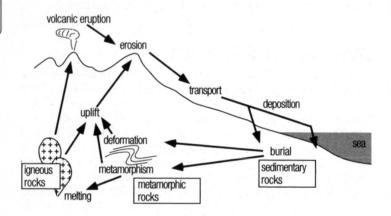

figure 1.1 cross-section through the Earth's surface to show the rock cycle.

In nature the rock cycle is not as simple as it may seem in Figure 1.1. If we were to attempt to trace the history of a particular volume of rock over hundreds of millions of years, we might find it tracing a loop to and fro through just part of the cycle, perhaps being alternately metamorphosed and melted at depth and never being brought to the surface. On the other hand, a grain of sediment may be deposited in a whole succession of sedimentary environments, being plucked out by erosion and transported to each new environment without ever becoming deeply buried.

The rock cycle is the most important unifying theme in geology, and forms the main theme of this book. However, Earth scientists recognize many other interrelated cycles too. For example, the water that does much of the erosion, that transports and shapes sedimentary grains, and that carries in solution the chemicals that help to turn soft sediment into hard sedimentary rock is part of a **hydrological cycle** involving evaporation and precipitation, interaction with the ocean floor, and even passage into the Earth's interior and escape again during volcanic eruptions. Similarly there is a **carbon cycle**, involving an interplay between atmospheric carbon dioxide and carbon stored short-term in living organisms and long-term in a certain kind of sedimentary rock (carbonates, such as limestone) and in coal and oil. This is an important influence of climate, because the more carbon dioxide there is in the atmosphere, the more solar heat is trapped and the warmer the planet tends to be.

Structure of this book

There is no obvious point at which to begin describing the rock cycle. In this book I begin with the igneous parts of the rock cycle, near the surface (Chapter 05) and at depth (Chapter 06). Next I deal with metamorphism (Chapter 07), followed by the erosional and depositional parts of the cycle (Chapters 08 and 09), and then the remaining aspect of the rock cycle, which is how rocks are deformed (Chapter 10). Subsequent chapters consider the physical resources that are available to us as a result of the rock cycle, the history of life on Earth and of the Earth itself, and evidence that the processes we have described occur on other planets too. The final chapter provides an introduction to geological fieldwork.

First, though, I describe the Earth's internal properties (Chapter 02), earthquakes (Chapter 03), which are one of the most obvious manifestations of the unquiet Earth, and plate tectonics (Chapter 04), which is a description of how wandering continents and widening or vanishing oceans are associated with the creation and destruction of rock on a huge scale.

02

the solid Earth

In this chapter you will learn:
- about the Earth in cross-section, from the atmosphere to the core, and how some of this information has been obtained
- about the Earth's age and origin.

Before exploring the geological processes that affect the Earth, we had better establish just what we are dealing with. So, what is the Earth? As Table 2.1 shows, it is the largest of several rocky planets in our Solar System. It is nearly spherical, measuring 12 714 km from pole to pole and 12 756 km across the Equator, and orbits the Sun at an average distance of 150 million km. We will look at these and other geologically interesting bodies elsewhere in the Solar System in Chapter 14, but until then our focus is very much on the Earth.

Name	Distance from Sun in millions of km	Equatorial radius in km	Mass, relative to Earth (6.0 x 10^{24} kg)	Density in tonnes per cubic metre (10^3 kg m^{-3})	Surface atmospheric pressure (bars)
Mercury	58	2439	0.055	5.43	negligible
Venus	108	6052	0.815	5.25	92
Earth	150	6378	1	5.52	1
Moon	150	1738	0.0123	3.34	negligible
Mars	228	3394	0.107	3.90	0.0016

table 2.1 The Earth and similar planetary bodies. Although the Moon orbits the Earth rather than the Sun it is included here because geologically it belongs to the same class of bodies.

The atmosphere and its development

Compared with the size of the planet, Earth's atmosphere is no more than a very thin layer. It is so sparse as to be virtually unbreathable at the tops of the tallest mountains (8–9 km above sea level), the clouds rarely occur higher than at about 12 km, and it is insubstantial enough to offer little or no resistance to artificial satellites orbiting at as low as 200 km altitude. However, so far as living things are concerned, the importance of the atmosphere is out of all proportion to its size. This is not just because it contains the oxygen necessary for breathing, but also because the atmosphere moderates the temperature of the globe and shields the surface from radiation that would be harmful to life. Most familiar living things would have a hard time surviving on the Earth's surface if the atmosphere were to become radically changed.

This is not to say that the present atmosphere has been around since the Earth was formed. As you will see towards the end of this chapter, the Earth's origin is reliably dated at about 4.57 billion years ago. At that time the Earth's primordial atmosphere, most of which is thought to have escaped from the interior by means of volcanoes, was probably mostly water vapour, carbon dioxide, sulphur dioxide and nitrogen. The appearance and subsequent evolution of exceedingly primitive living organisms (bacteria-like microbes and simple single-celled plants) began to change the atmosphere, liberating oxygen as carbon dioxide and sulphur dioxide were broken down. This, in turn, made it possible for higher organisms to arise.

When the earliest known plant cells with nuclei appeared, about 2 billion years ago, the atmosphere seems to have had only about one per cent of its present content of oxygen. With the arrival of the first land plants, about 500 million years ago, oxygen reached about one-third of its present concentration. It had risen to almost its present level by about 370 million years ago, when animals first emerged onto land. Today's atmosphere is thus not just a requirement to sustain life as we know it; it is also a consequence of life.

The Earth's interior

This book is concerned mostly with things that occur upon, or at shallow depth below, the Earth's surface. We will come back to the atmosphere from time to time, but first we need to investigate the deep interior. People living in regions where the soil is thin are well aware that the outside of the solid Earth is made of rock. The rest of us are reminded of this when we see a quarry or a cliff. But could the Earth be made of rocky material like this all the way to the centre?

A simple way to determine if that is possible is to consider the Earth's density. The kind of rock we usually find at the surface has a density of about 2.7 tonnes per cubic metre. If the Earth were made of rock, it should have an overall density only a little greater than this value, to allow for compression, and hence greater density, at depth.

We can work out the Earth's actual density as follows. The Earth's mass is known to a high degree of precision, because it can be determined from how long it takes the Moon or an artificial satellite to complete an orbit. It is shown in Table 2.1 as 6.0×10^{24} kg. If you are not familiar with scientific notation, this translates as 6 million million million million kilogrammes, or 6 million billion billion kilogrammes, or 6 thousand billion billion tonnes. To find the Earth's density, we just divide its mass by its volume. The result is 5.52 tonnes per cubic metre, which is about twice the density of surface rock. Clearly, therefore the Earth is not just rock. Since we can see that the outer part is rock, we can deduce that there must be something denser than the average value hidden away at depth. In fact, the centre of the Earth is believed to be composed mostly of iron, with a density of almost 13 tonnes per cubic metre.

The picture geologists have formed of the Earth's interior is not based merely on density arguments. All that density tells us is that there must be something much denser than rock somewhere within the Earth. It does not tell us how this denser matter is arranged, let alone what it is made of. For example, Figure 2.1 shows two entirely different types of internal structure for the Earth, both of which are consistent with the Earth's density, but only one of which makes sense when other evidence is taken into account.

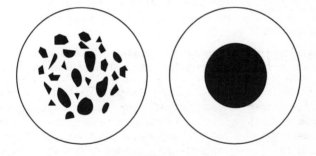

figure 2.1 Two possible arrangements of dense matter within the Earth. Either of these would account for the observed average density, but only the model on the right fits the observations discussed later in the text.

Seismic waves

The most useful information for determining the internal structure of the Earth comes from the study of how vibrations travel through it. As you can imagine, it needs pretty powerful vibrations to be detected through the full thickness of the globe. Fortunately, the vibrations from large **earthquakes**, known as seismic waves, are sufficiently strong for this purpose (though this is a mixed blessing for people living in the most earthquake-prone areas, as you will see in the next chapter).

The reason why seismic waves are so useful for probing the Earth's interior is that the speed at which vibrations travel through a substance depends on its physical properties. As density increases, so the speed of seismic waves tends to decrease. On the other hand, vibrations travel faster through a more rigid substance than through a less rigid substance.

Let's see how this can help us determine the structure of the Earth, by supposing for now that the outer part of the Earth has a uniform (rocky) composition. How would we expect the speed of seismic waves to vary with depth? Rigidity should increase with depth, as a result of the ever-increasing pressure exerted by the overlying material. The effect of this should be for the speed of transmission of seismic waves to increase as depth increases. The contrary tendency for speed to decrease because of increasing density with depth is a much slighter effect, since rock is not compressed very much however high the pressure gets. We therefore expect the speed of seismic waves to increase with depth within the rocky part of the Earth, and this is indeed what we find.

We can tell this when an earthquake happens at a known location by timing the arrivals of the first vibrations received by

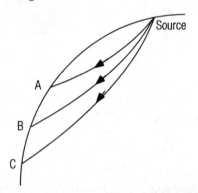

figure 2.2 Because the speed of transmission of seismic waves increases with depth, the first vibrations detected by distant seismometers are not those that have travelled in a straight line, but those that have followed a curved track, shown here, taking them on a deeper but faster route. The first vibrations reach seismometer A about 27 minutes after the event, and seismometer C only about 35 minutes after the event, even though C is about one and a half times further away.

sensitive vibration detectors, known as **seismometers**, at different distances. Figure 2.2 shows how this works. Note that the vibrations follow curved paths as a consequence of the speed of the transmission increasing with depth.

Studies of this sort, based on hundreds of thousands of earthquakes detected at thousands of seismometers across the globe, has enabled the speed of seismic waves at different depths to be determined in considerable detail. The structure of the Earth's outer few tens of kilometres varies considerably from place to place, and we will consider this shortly. The deeper interior is much more uniform, seismic speeds changing with depth but not varying much with location.

What helps us to pin down the density distribution within the Earth is that at a depth of 2900 km the speed of seismic waves drops sharply. We can tell this because instead of continuing to follow an upward-curving track, the path of a vibration reaching this depth is deflected *downward*, as shown in Figure 2.3. At this depth the composition must change from rocky to

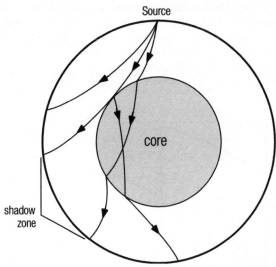

figure 2.3 Paths of seismic waves deep within the Earth. The density of the core is about twice that of the much more rigid overlying mantle, so seismic speed is halved. The direction of wave travel is deflected sharply downward at the core–mantle boundary, and there is a 'shadow zone' at a certain distance from the source within which seismometers cannot pick up directly transmitted waves. In addition to earthquakes, underground nuclear explosions cause seismic waves powerful enough to probe the deep structure in this way.

something different, with about twice the density, causing a dramatic decrease in seismic wave speed. This marks the outer limits of the Earth's **core**. The rocky material surrounding the core is known as the **mantle**.

The Earth's core

A remarkable thing about the core at this depth is that, despite the immense pressure, it is not solid. This has been demonstrated by studying the transmission of different kinds of seismic waves, which so far we have glossed over. There are two sorts of seismic wave that concern us when considering the Earth's interior: compressional waves and shearing waves. A compressional wave (or P-wave) is just like a sound wave in air; it consists of alternate pulses of compression and dilation. A shearing wave (or S-wave) can be observed by shaking a jelly; it is an alternate side-to-side wobble travelling through the body of the material. An important distinction between these two types of seismic waves is that whereas compressional waves can travel through anything, a liquid cannot transmit shearing waves. This is because a liquid offers no resistance to shearing motion. Seismometers can distinguish between the two types of waves, and from this we can tell that only P-waves are transmitted through the outer core, which must therefore be liquid.

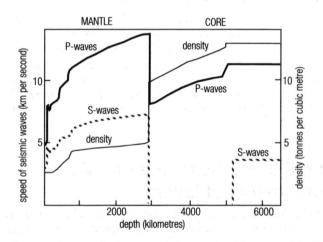

figure 2.4 Speeds of compressional waves (P-waves) and shearing waves (S-waves) within the Earth, and inferred density.

That is not quite the end of the story, because we can identify an inner core whose limit is marked by a rapid increase in the speed of P-waves and within which we can tell (by indirect means) that S-waves are once again transmitted. Figure 2.4 is a plot showing the variation with depth of the speeds of P-waves and S-waves (which can be calculated) and density (which can then be deduced).

Now we have assembled the evidence needed to decide between the alternative models of density distribution given in Figure 2.1. We can see that the right-hand model (symmetrical distribution), rather than the left-hand model (irregular distribution), is correct. In fact, we can go further than this, because we have enough information to propose what the dense stuff actually is. The solid inner core has the properties we would expect of solid iron mixed with a few per cent nickel. Its properties match that of other metals too, such as cobalt or titanium, but the iron core model is far more reasonable because it fits with the Earth being rich in the same metallic elements as we find in the Sun and in meteorites. The liquid outer core has a density too low to be pure metal. About ten per cent of its mass must be composed of one or more relatively light elements. What these are cannot be proved, but oxygen, sulphur, carbon, hydrogen and potassium (a light metal) are the most likely.

Evidence from magnetism

There is one more factor bearing on the nature of the core that we should touch on here, and this is magnetism. The Earth has a magnetic field resembling one produced by a giant bar magnet (Figure 2.5). Both common sense and the seismic evidence we have just investigated enable us to rule out a real bar magnet as the origin of this magnetic field. Instead, the field is regarded as the product of electrical currents generated by motion within the fluid outer core.

The Earth's crust and mantle – compositional layers

Now we will turn to the outermost few tens of kilometres of the Earth, where the variation in seismic speeds is most complex. Junctions between one rock body and another are sometimes manifested by sharp changes in seismic speed, but on a global scale the effect is one of gradually increasing speed with depth.

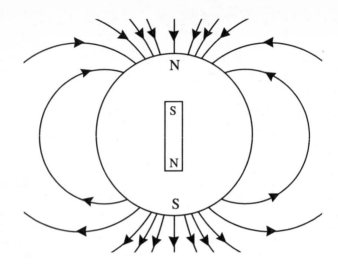

figure 2.5 Cross-section through the Earth from pole to pole, showing the lines of magnetic field and an imaginary bar magnet at the Earth's centre.

This is because of the dominating effect of increasing rigidity when depth, and hence pressure, increases. P-wave speed increases gradually from about 2 km per second just below the surface to 6 or 7 km per second. Below this, a sharp jump to a P-wave speed of 8 km per second is recognized throughout the entire globe, at an average of 30 km below the continents but usually 10 km or less below the ocean floor. This sharp change to denser rocks is known as the Mohorovicic discontinuity (after Andrija Mohorovicic, 1857–1936, the Croatian seismologist who first recognized it). It is usually called the **Moho** for short. Above the Moho are the rocks that belong to the Earth's **crust**, and below that we are into the mantle, which extends all the way to the core.

The crust is composed of slightly less dense rock than the mantle, having a greater proportion of silicon, aluminium, calcium, sodium and potassium, but less magnesium. It is easy to explain why the crust sits above the mantle; essentially it is just the light stuff that has worked its way to the top. For reasons that you will learn later, there are actually two distinct types of crust: continental crust (underlying virtually all the land surface and the shallow seas) and oceanic crust (forming the floor of the deep oceans). The compositions of these two types of crust are compared alongside the mantle composition in Table 2.2.

element	oxide	Continental crust (granite-andesite)	Oceanic crust (basalt)	Mantle (peridotite)
silicon	SiO_2	62	49	45
titanium	TiO_2	0.8	1.4	0.2
aluminium	Al_2O_3	16	16	3.3
iron	Fe_2O_3	2.6	2.2	1.2
iron	FeO	3.9	7.2	6.7
magnesium	MgO	3.1	8.5	38.1
calcium	CaO	5.7	11.1	3.1
sodium	Na_2O	3.1	2.7	0.4
potassium	K_2O	2.9	0.26	0.03

table 2.2 Average compositions of continental crust, oceanic crust and the mantle. The names of the rock types most closely matching these compositions are shown. Elements are named in the first column, but compositions are expressed as the oxides of these elements (per cent by weight) whose chemical formula is shown in the second column; this is purely a convention and should not be thought of as indicating the chemical species actually present. Iron is listed twice, as Fe_2O_3 and FeO. SiO_2 is commonly referred to as silica.

The thickness of the continental crust varies from about 25 km in thin, stretched regions to as much as 90 km below the highest mountain ranges. Oceanic crust is much thinner, ranging from about 6–11 km thick.

The wide variation in the thickness of continental crust is manifested not so much by mountains tens of kilometres high (the highest mountain, Everest, reaches only 8848 m above sea level), as by enormous downward protruberances of the base of the crust, which we can detect as variations in the depth of the Moho. This shows that mountains are not held up by the *strength* of the material on which they rest. Instead, crustal regions of different thickness may be thought of as floating on the denser mantle. The situation is illustrated in Figure 2.6. Remember, though, that the mantle cannot really be a liquid, because it transmits S-waves. It just behaves like a liquid when considered over geological time.

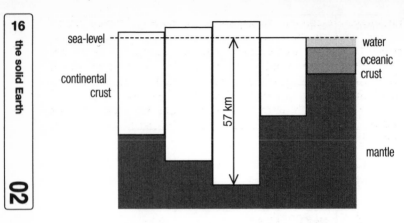

figure 2.6 Cross-sections showing crust of different thicknesses resting in equilibrium upon the mantle. In the situation shown, the weights of all five columns (crust plus mantle, and including water in the right-hand column) from surface to base are equal. This demonstrates how mountains of modest height (6 km in this example) are supported by buoyant roots (extending 57 km below sea level in this example). Average densities of water, continental crust, oceanic crust and mantle are 1.0, 2.7, 3.0 and 3.3 tonnes per cubic metre, respectively.

The concept of blocks of crust floating in equilibrium upon the mantle is known as **isostasy,** and has been found to apply almost everywhere. There are few examples known where high or low regions are held up or pushed down by any force other than simple buoyancy (which is really what is meant by isostasy). Buoyancy shows us why oceanic crust is virtually never found on dry land. It is thinner and denser than continental crust, and so it floats lower.

Summary of compositional layers

We can summarize what we have said so far about the Earth's compositional layers as shown in Figure 2.7. The compositional difference between crust and mantle is relatively slight, but sufficient to account for the jump in the speed of seismic waves across the Moho. The most abundant elements in both are silicon and oxygen. Any compound made of a chemical combination of just these two elements is known as **silica**. The rocky materials dominated by silicon and oxygen are therefore commonly referred to as **silicates**. Some depth-related changes in seismic speeds within the mantle have been recognized (see Figure 2.4), but these are thought to represent pressures at which atoms within crystals become packed into denser, more rigid, structures, rather than being changes in chemical composition. Throughout, the mantle is thought to have a chemical composition similar to that of the rock type known as **peridotite**. The compositional distinction between the core and the mantle is much more fundamental than between the crust and the mantle. The core does not even consist of silicates, but is dominated by iron.

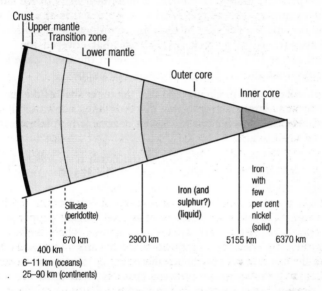

figure 2.7 The compositional layers within the Earth. The lower mantle has a denser structure than the upper mantle, but is not thought to differ significantly in composition.

Another kind of layering – strong and weak

When we consider how the Earth behaves, it is not always chemical composition that is the most important factor. We also need to bear in mind how strong the material is. We have already met this concept when we considered the crust as blocks 'floating' in the mantle. Mountain belts need deep buoyant roots only because the crust and underlying mantle do not have the strength to support them. Actually, the weakest zone in the mantle is not immediately below the crust, as you might think from a simple interpretation of Figure 2.6. Instead it begins at a depth of about one hundred kilometres. At this depth there is a slight drop in seismic speeds (discernible in Figure 2.4), which has been interpreted as due to the presence of a few per cent of molten material. Evidently, though, the material as a whole is still well and truly solid because S-waves can travel through it. This is known as the 'low speed layer' (sometimes the 'low velocity layer') and is weak enough to deform (over geological time periods) under the weight of mountain belts or ice sheets. For example, Scandinavia is still rising upwards at a rate of a few millimetres per year in response to unloading caused by the disappearance of a thick sheet of ice that covered the region until about 10 000 years ago. It is flow in the weak zone in the mantle that allows this slow rebound to happen.

You will discover in Chapter 04 that the outer shell of the Earth is broken into large slabs or 'plates' that are gliding sideways. This too would not be possible but for this weak zone, which lubricates the bases of the plates.

In terms of how the outer part of the Earth responds to forces, the crust and the uppermost part of the mantle constitute a single rigid unit known as the **lithosphere** (from *lithos*, the Greek word for rock). The lithosphere is rocky not just in its composition but also in terms of its mechanical properties. In contrast, below the lithosphere the mantle is relatively weak, although its chemical composition is the same. This weak zone has been named the **asthenosphere** (using the Greek word for weak). The relationship between crust, mantle, lithosphere and asthenosphere is summarized in Figure 2.8.

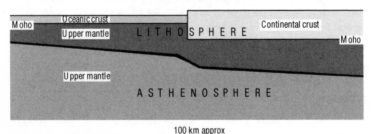

100 km approx

figure 2.8 The mechanically distinct outer layers of the Earth; the rigid lithosphere and the relatively weak asthenosphere.

The weakest part of the mantle lies in the few tens of kilometres immediately below the base of the lithosphere, and the term asthenosphere is often restricted to just this thin layer (coinciding with the low speed layer). However, once we get below the lithosphere there has been an important change in the properties of the rock that persists all the way to the core. Although solid in respect of the transmission of seismic waves (which travel at many kilometres per second), the rock below the lithosphere is not at rest. It is circulating slowly, at speeds of a few centimetres a year. This is usually described as 'solid-state convection'. Convection is what makes warm air rise and cold air sink, or water circulate in a saucepan (even before it boils). It is a way of transporting heat outward. Put simply, hot mantle rises upwards and transfers its heat to the base of the lithosphere. Circulating mantle that has lost heat in this way becomes slightly denser, and sinks downwards to complete the loop (Figure 2.9, Plate 1). Most of the heat deposited at the base of the lithosphere trickles through to the surface by conduction, but some is carried higher by molten bodies of rock that can intrude high into the crust, or even reach the surface at **volcanoes.**

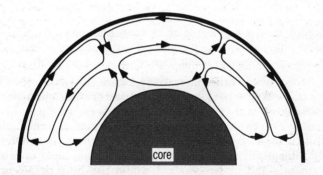

core

figure 2.9 Possible arrangement of convection within the sub-lithospheric mantle. This model shows two layers of convection cells, meeting at the seismically defined transition zone (see Figure 2.7).

The Earth's heat

Considering the nature of the mantle has brought us to the question of the Earth's heat. It is common knowledge that the interior of the Earth is hot, but most people probably think this means that below a certain depth everything is molten. We have already seen that such a notion is misconceived, because seismic data prove conclusively that the mantle – even the bulk of it that constitutes the asthenosphere – is virtually all solid.

The mantle *is* hot, but not as hot as you would think by extrapolating from surface measurements. Heat is escaping to the surface at a rate of about eight-hundredths of a watt per square metre (0.08 W m^{-2}), so you would need to collect all the heat escaping from an area greater than 1000 m^2 to power a 100 watt light bulb. Near the surface of continents, the average rate of temperature rise with depth (the **geothermal gradient**) is about 30 °C per kilometre. From this we could infer, by extrapolation, a temperature of about 900 °C at the base of 30 km thick crust and about 2700 °C at the base of 90 km thick crust, but we would be wrong to do so. Certainly temperatures over 1000 °C are not reached, because crustal rocks would melt at about this temperature.

The solution to this conundrum is that only about half the heat escaping to the surface of the continents comes from below, whereas the rest is actually generated within the crust. This means that the geothermal gradient becomes less as you go deeper.

If fact, below the lithosphere, temperature rises at a rate of less than half a degree per kilometre. One factor contributing to this low value is that, as we have already seen, this part of the mantle is convecting the heat. Convection is much more effective than conduction at transporting heat. If convection could (magically) be stopped, then heat would build up within the mantle and its temperature would rise by a degree or so every million years until it melted. The action of solid-state convection within the mantle is actually what stops it becoming hot enough to melt.

The very effective convection in the liquid outer core keeps the temperature gradient even lower there, so that the temperature of the inner core is estimated at a surprisingly low value of approximately 4700 °C.

So where does the Earth's heat come from? An unknown proportion, perhaps around 30 per cent, is heat that was trapped within the Earth as it formed. This primordial heat continues to leak out, but is not being generated today. The rest of the heat comes from processes that are continuing today. By far the greatest of these is decay of radioactive elements, leading to so-called **radiogenic heating**.

Radiogenic heating

Many elements have radioactive isotopes, but only three of these are important heat sources in the Earth today. These are potassium, uranium and thorium. Most potassium atoms are stable and non-radioactive. Each of these contains a total of 39 heavy particles (protons and neutrons) in its nucleus, and is described as potassium-39, or ^{39}K (using the chemical symbol for potassium, K). However, about one potassium atom in every 10 000 contains an extra neutron, making a total of 40 heavy particles. These atoms of potassium-40 (^{40}K) are unstable, and undergo radioactive decay. All atoms of uranium and thorium are unstable; thorium occurs as ^{232}Th, and uranium has two radioactive isotopes, ^{235}U and ^{238}U.

Although it is impossible to predict when any individual radioactive atom will decay, there is a measurable probability of it happening during a given time. When dealing with large numbers of nuclei, probability takes on a deadly certainty. We can therefore determine the time it takes for half the nuclei of a particular isotope to decay, which is known as the half-life. Because the number of radioactive nuclei halves over a fixed time, it never actually reaches zero (although it can get very close to it for isotopes with much shorter half-lives than those discussed here). The half-life of ^{238}U is 4.5 billion years. This, coincidentally is roughly the age of the Earth, so it means that the abundance of ^{238}U in the Earth must now be very close to half what it was when the Earth formed. The half-life of ^{232}Th is over three times the age of the Earth, so about 80 per cent of what the Earth started with is still there.

Of the four heat-producing isotopes, ^{40}K (half-life 1.3 billion years) probably produces the greatest amount of heat today and ^{235}U the least. However, ^{235}U has the shortest half-life of the four, about 710 million years, which is about one-sixth the Earth's age. To determine the rate of heat production by ^{235}U when the Earth was very young we must multiply the present rate by 2^6 or 64 times, from which we can deduce that it was the second most important heat-producing isotope during the Earth's first half billion years.

All three of these elements tend to be concentrated into the relatively light rocks of the continental crust, rather than the mantle, which explains why as much as half of the Earth's radiogenic (radioactively produced) heat is generated in the crust, although this makes up less than 0.1 per cent of the Earth's total volume. However, one of the great unknowns is whether potassium is abundant in the outer core, where it is a contender for the unknown light element. What is not in any doubt at all is that the Earth's radioactive power source must be running down, though at a rate that is too slow to concern us for most purposes.

Radiometric dating and the age of the Earth

Apart from heat production, radioactive decay is important to the geologist because it is a physical process whose rate we know. The time since a rock or crystal was formed can be determined by measuring the relative proportions of the surviving parent isotope and its decay products trapped within the same rock or crystal. This is called **radiometric dating**. For example, ^{40}K (the parent isotope) decays to ^{40}Ar (its decay product). Argon is not commonly trapped within minerals when they are formed, so it is usually assumed that any argon found within a mineral is the product of decay of ^{40}K. Unfortunately argon can escape from within a crystal, so potassium-argon dates are sometimes unreliable.

The decay of uranium (to lead) offers an important dating technique, but because of its extremely long half-life ^{232}Th is not used much. There are other radioactive isotopes that are insignificant producers of heat but whose decay products enable radiometric dating to be performed. The most important of these is the decay of rubidium-87 (^{87}Rb) to strontium-87 (^{87}Sr). Radiometric dating techniques can be used to determine the time since a crystal grew from molten rock or by recrystallization within rock without melting (a process known as metamorphism). Extensions of these techniques have enabled the age of the Earth, and of ordinary meteorites, to be determined at 4.57 billion years with a fairly high degree of confidence.

The origin of the Earth and the Solar System

Now that we have completed our brief survey of the Earth's interior and gross composition, it is time to describe how scientists think the Earth came into being. This takes us back to the birth of the Sun, which grew by the collapse of a slowly rotating interstellar cloud of gas (mostly hydrogen) and dust. This cloud is referred to as the solar nebula. As it contracted, the solar nebula spun faster and faster. The material that was not drawn into the central point, where the Sun was forming, became concentrated in a disc around the Sun. It was within this disc that the planets grew. At first the cloud would have been very hot, because of the gravitational energy converted to heat by the contraction. As it cooled, things began to condense out of the gas as tiny solid grains. The first grains to condense would have been made of substances able to form at temperatures above 1000 °C. These would have included nickel-iron metal and some of the silicate minerals. As the temperature declined to a couple of hundred degrees centigrade the minerals growing could begin to trap water within their structure, but ice crystals would not begin to grow until the temperature reached –90 °C.

The tiny grains that formed within the solar nebula would tend to stick together whenever they happened to come into contact, and so progressively larger chunks would collect. Once the process had begun, it may have taken as little as a few thousand years to form centimetre-sized pieces. Throughout the sequence of events the timescales are poorly understood, but the following is a reasonable summary. After about 100 thousand years, the biggest blocks had probably grown to about 10 km across and are dignified with the name 'planetesimals'. These were now big enough to make their gravitational influence felt, and their growth would now proceed at an accelerating rate until after about a further 50 thousand years most of the planetesimals had collided and accreted into a few dozen bodies a few thousand kilometres across, now known as 'planetary embryos'.

Collisions between these surviving bodies would have been violent, sometimes shattering both bodies, but more often pasting the debris of the smaller body across the face of the larger one and liberating sufficient heat to cause melting to a very great depth. After about 100 million years all the planetary embryos that were going to collide had probably done so, leaving the Solar System with four inner rocky planets (see Table 2.1). There were also four giant planets (Jupiter, Saturn, Uranus and Neptune) further out from the Sun where it had become cold enough for large quantities of water, methane, ammonia and similar volatile substances to condense, and relatively tiny icy objects (Pluto being the most famous but no longer the largest known) in the outer fringes where the density of material was too sparse to collect into large bodies.

The Earth probably acquired the Moon during these latter stages, forming from the debris of the last planetary embryo collision to affect the Earth. In this case, some fragments of the impacting body bounced back from the Earth and collected together in orbit around the Earth.

The irregularly shaped rocky and iron-rich bodies known as the asteroids, which are up to a few hundred kilometres across and found mostly between the orbits of Mars and Jupiter, were once thought to be the remains of a planet that was broken up by a giant collision. However, it now seems that they are surviving planetesimals that never stuck together when they hit each other, because their orbits were stirred up by their proximity to Jupiter so that they hit each other too hard to allow accretion. Most meteorites, which are chunks of rock or nickel-iron that fall from the sky, are thought to be made of the same material as the asteroids.

A warning

The origin story as outlined above may sound rather glib. It is important to realize that we are on the threshold of the realm of legend here. Most scientists would accept this story in outline, but there are many important details still in dispute. There is good evidence that the majority of Sun-like stars appear to have their own planetary systems. Curiously, most of the documented 'exoplanets' are in orbits very close to their stars, which has earned them the apt name of 'hot Jupiters'. Such 'solar systems' are clearly unlike our own, and it is believed that giant exoplanets such as those must have formed further away from their star and then migrated inwards, a process that would almost certainly destroy any intervening Earth-like exoplanet. However, these examples are probably unrepresentative, because giant exoplanets close to their stars are the easiest ones to find. The first Earth-sized exoplanet with a surface cool enough for liquid water was discovered as recently as 2007, and there are hopes that within a further decade or two we may be able to determine at least the atmospheric composition of such worlds, even though details of their geology must remain a mystery.

Closer to home, the temperature history of the solar nebula is uncertain. It could have remained too hot for much rocky material to have condensed until substantial sized bodies of nickel-iron had formed. In this case, the later-formed rock would have collected around these pre-grown cores and planetary bodies would develop with an in-built layered structure. On the other hand, the temperature may have fallen swiftly enough for chunks of a wide range of compositions to have been around at the same time. Planetary bodies would then have grown with a thoroughly mixed composition, and internal layering would then have to be generated by some later process.

It is not too difficult to envisage how compositional layering could develop from initially well-mixed material, especially if a planet like the Earth grew by collisions between planetary embryos of similar size. As noted above, these impacts would have been violent enough to cause global melting, and this would have provided the opportunity for dense components, such as iron, to sink and form the core, while lighter materials, such as the silica-rich minerals characteristic of the crust and mantle, rose upwards. However, a variant of the above story omits the planetary embryo stage, and has each planet developing from the largest planetesimal in its vicinity, growing by accreting other planetesimals that were always much smaller than itself. In this scenario, the heat released by impact energy may not have been enough to cause global melting, in which case differentiation into a compositionally layered structure would have to depend on internal sources of heat, such as radioactive decay.

We will put speculation aside now, and turn to much firmer ground (metaphorically speaking!) and look at what happens in an earthquake.

03

earthquakes

In this chapter you will learn:
- about earthquakes and how they are measured and predicted
- about the dangers posed by earthquakes, and how risks can be minimized.

You have already seen that earthquakes are useful for probing the interior of the Earth. This is because the seismic waves generated by earthquakes can travel right through the Earth. However, earthquakes themselves do not happen deep down, but are pretty much confined to the lithosphere. It is their very shallowness that makes the larger ones so devastating. In this chapter we will concentrate on describing earthquakes. Their fundamental causes will become apparent in the next chapter.

How earthquakes happen

There are many fault lines in the Earth's crust. The San Andreas fault in California is probably the most famous. A **fault** is a fracture between tracts of crust that are moving relative to one another. The typical average rate is around a millimetre per year. If this movement were to happen gradually it would pose few problems for people living nearby. Unfortunately rocks do not behave that way. Instead, they tend to stick. Strain builds up for decades or centuries until it reaches a critical level, and then everything gives at once. The principle is shown in Figure 3.1. Something similar to what goes on at the break-point of an earthquake happens if you stack two house bricks on top of each other. If you hold the lower one, and gradually tilt it you will find that the upper one does not move at first. As the angle of tilt increases, the force tending to make the top brick slide gradually increases until it is great enough to overcome friction, and suddenly it will slide off.

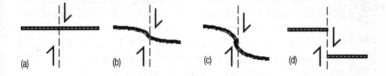

figure 3.1 Plan view showing a road built across a fault (dashed line). Strain builds up gradually, (a)–(c), as the edges of the rock-masses on either side of the fault are bent elastically. Eventually, (d), the strain becomes great enough to overcome the frictional resistance along the fault and the fault gives way. Slip on the fault is rapid, and releases a lot of pent-up energy as seismic waves.

As soon as a fault has given way at one point, slip movement may occur along its whole length, which may be hundreds or thousands of kilometres although, other than in exceptionally large earthquakes, movement is usually restricted to a much shorter portion of the whole fault. The strongest seismic waves are generated at the initial break-point. The closer to the break-point, the greater the energy and the greater the potential for destruction, especially if the break-point is near the surface. Technically, the break-point is called the earthquake **focus**, and the point on the surface directly above it is referred to by that much-abused term **epicentre**. Slip further along the fault and readjustments close to the focus usually cause a series of smaller aftershocks, which continue for days (even years in extreme cases) after the initial earthquake.

Little, if any, of the damage caused by an earthquake is done by the P-waves and S-waves that we considered in the previous chapter. These waves travel through the body of the Earth, and their energy is spread over a rapidly increasing volume as they propagate. It is other waves of the sort that travel along the surface of the ground that do most of the harm. These include up-and-down waves (like waves on the ocean, called Love waves) and strong side-to-side or to-and-fro shaking waves (called Rayleigh waves). The P- and S-body waves travel faster than the surface waves, and, if felt, can give a few seconds (minutes, if further away) warning of the arrival of the more damaging surface waves.

Measuring earthquakes

When an earthquake makes the news, its size is usually described by a number on the Richter scale. This scale (of which there are several variants) is an attempt to specify how strong the seismic waves must have been at the focus. This can be estimated using the signal detected by a single seismometer at a known distance from the focus or, better, by comparing the signals detected at many seismometers at a variety of distances. The Richter scale is numerical, enabling comparative data to be compiled on earthquakes of all sizes from around the globe. For each one magnitude increase on the scale, the motion (or amplitude) of the seismic waves at the focus increases tenfold and the energy released increases roughly thirtyfold. The most devastating earthquakes tend to be of magnitude 6.0 and above. Fortunately there is on average only one earthquake bigger than magnitude 8.0 each year. Table 3.1 summarizes the occurrence of earthquakes on the Richter scale.

Magnitude range	Number per year	Example (number of deaths in brackets)
8.0 and above	1	San Francisco 1906 M8.3 (500) Tokyo 1923 M8.2 (143 000) Kurile Islands 1994 M8.2 (12) Sunda Trench 2004 M9.1 (283 000) Sunda Trench 2005 M8.7 (1313) Coast of Peru 2007 M8.0 (514)
7.0–7.9	18	Tangshan, China 1976 M7.6 (250 000) Mexico City 1985 M7.9 (30 000) Loma Prieta (San Francisco) 1989 M7.1 (61) East Cape Town, North Island, New Zealand 1995 M7.5 (none) Kobe, Japan 1995 M7.2 (5000) Izmit, Turkey, 1999 M7.6 (17 118) Bhuj, India 2001 M7.7 (20 000) Muzzafarabad, Kashmir 2005 M7.6 (80 000)
6.0–6.9	134	Avellino, Naples, Italy 1980 M6.8 (8000) Spitak, Armenia 1988 M6.8 (25 000) Northridge (Los Angeles) 1989 M6.7 (61) Bam, Iran 2003 M6.6 (26 200) North Island, New Zealand 2007 M6.6 (1)
5.0–5.9	1300	Plattsburgh, New York 2002 M5.1 (none) San Giuliano di Puglia, Italy 2002 M5.3 (29) Market Rasen, England 2008 M5.2 (none)
4.0–4.9	13,200	Coalinga, California 1983 M4.3 (none) Dudley, England 2002 M4.8 (none) Kent, England 2007 M4.3 (none)
3.0–3.9	130,000	Fort William, Scotland 2005 M3.0 (none)
2.0–2.9	1,300 000	
<2.0	millions	

table 3.1 Numbers of earthquakes per year in each magnitude range of the Richter scale. Some well-known, or typical, earthquakes of each magnitude are listed.

A disadvantage of the Richter scale is that it does not show how much damage a particular earthquake is capable of. For example, deep focus earthquakes are not very effective at generating surface waves, so they tend not to cause much damage. Furthermore, what effect an earthquake has depends on when and where it occurs. A shallow magnitude 8.0 earthquake in the middle of nowhere will probably cause much less damage than a shallow magnitude 6.0 earthquake whose epicentre is in a major city, especially if it happens during the rush hour. The effect of an earthquake also depends on the nature of the rock near the surface. In the Mexico City earthquake of 1985 the soft sandy subsoil quivered so vigorously that some buildings sank deep into it, causing many deaths. A similar earthquake below a city built on hard rock would probably have caused much less damage.

An alternative to the Richter scale is the Mercalli scale. This ranks earthquakes according to the intensity of their effects, and can be used to map the extent to which concentric zones around the epicentre are affected by a single earthquake. A version of it is shown in Table 3.2.

Some of the descriptions on the Mercalli scale are universally applicable, such as whether the earthquake can be felt by people who are walking about, or only by those lying down. However, the amount of damage to buildings depends very much on the nature of the construction. A well-designed structure can withstand even a very major earthquake without collapsing, even though it may be so badly damaged that the only safe thing to do afterwards is to demolish it. Most earthquake-prone regions have building codes designed to ensure that structures are relatively collapse-proof, and failure to observe and enforce these codes has led to many unnecessary deaths.

Mercalli intensity	Description
XII	Damage total. Lines of sight and level distorted. Objects thrown in the air.
XI	Few if any masonry structures remain standing. Bridges destroyed. Rails bent greatly.
X	Some well-built wooden structures destroyed; most masonry and frame structures destroyed with foundations. Rails bent.

IX	Damage considerable in specially designed structures; well-designed frame structures thrown out of plumb. Damage great in substantial buildings, with partial collapse. Buildings shifted off foundations.
VIII	Damage slight in specially designed structures; considerable damage in ordinary substantial buildings, with partial collapse. Damage great in poorly-built structures. Fall of chimneys, factory stacks, columns, monuments, walls. Heavy furniture overturned.
VII	Damage negligible in buildings of good design and construction; slight to moderate damage in well-built ordinary structures; considerable damage in poorly-built or badly designed structures; some chimneys broken. Difficult to stand upright. Noticed by vehicle drivers.
VI	Felt by all, many frightened. Some heavy furniture moved; a few instances of fallen plaster. Damage slight.
V	Felt by everyone; many awakened. Some dishes and windows broken. Unstable objects overturned. Pendulum clocks may stop.
IV	Felt by many indoors, outdoors by few. At night, some awakened. Dishes, windows, doors disturbed; walls make cracking sound. Sensation like heavy truck striking building. Standing cars rock.
III	Felt quite noticeably by people indoors, especially on upper floors. Many people do not recognize it as an earthquake. Hanging objects swing. Feels like passing traffic.
II	Felt only by a few people at rest, especially on upper floors
I	Generally detected by instrument only.

Table 3.2 The Modified Mercalli scale of perceived earthquake intensity. The more intense earthquakes are categorized by structural damage, less intense earthquakes by effects perceived by people on the spot.

Protecting structures from earthquake damage

The are many ways to make buildings, elevated highways and other structures earthquake resistant. It is not simply a matter of making everything stronger. It is more important to ensure that

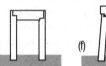

figure 3.2 Two series of cross-sections to show possible response of a simple structure (two walls and a roof) to the ground-shaking caused by an earthquake. In (a)–(d) both walls sway in the same direction simultaneously, and the structure survives with relatively minor damage. In (e)–(h) the walls sway in opposite directions, with catastrophic results. Here the roof collapses, even though the walls manage to stay upright. In inadequately designed multistorey buildings, several floors have been known to collapse on top of one another, squashing anyone unfortunate enough to have been inside.

if a structure is rocked by an earthquake, the whole of it sways together in the same direction (Figure 3.2).

As well as tying walls together so that they sway in unison, it is important to ensure that a building can sustain any twisting motion engendered by an earthquake. Digging extra-deep foundations can also help, so that the structure rests on solid bedrock where the amplitude of the vibrations is likely to be less, or at the least to make sure that the foundations rest on the same material all round so any subsidence happens evenly on all sides.

Another factor is the natural frequency at which a building tends to shake. Taller buildings sway more slowly than low-rise buildings, and there is a danger that high-rise buildings will be 'in tune' with the shaking of the ground. This had an unfortunate consequence in the major earthquake that struck Mexico City in 1985. In the most heavily damaged area of the city, about 70 per cent of buildings between 10 and 20 storeys high were damaged because the low-frequency shaking of the ground became amplified within the building by resonant vibrations of their structure. In contrast, fewer than 20 per cent of buildings fewer than five or more than 20 storeys high were damaged.

Searching for survivors after an earthquake is a risky business. Buildings that are apparently unscathed may in fact be dangerously near to collapse, which could be triggered by an aftershock at any time. There may also be a risk of explosions from ruptured gas pipes. Fire is actually one of the biggest hazards in the aftermath of a major earthquake. Most of the death and destruction associated with the San Francisco earthquake of 1906 was not a result of collapsing buildings but of fires, caused by toppled lamps and the like. The majority of the buildings survived the quake, but were razed to the ground by the subsequent fires.

Many lessons have been learned about how to build earthquake-resistant structures. However, even in the wealthier at-risk areas large numbers of old and less-resistant structures remain, as demonstrated by some of the damage caused by late twentieth-century earthquakes in California (Figure 3.3) and Japan.

figure 3.3 This building in San Fransicso was wrecked when its lower floors gave way during the 1989 Loma Prieta earthquake.

Earthquake preparedness

In earthquake-prone regions most people are well aware of the risks they face. Instructions on how to be prepared, and what to do in the event of an earthquake, are widely available and are commonly rehearsed in schools. The following is an extract from the advice pages of the Los Angeles telephone directory.

During an Earthquake

1 If you are indoors, **DUCK** or drop down to the floor. Take **COVER** under a sturdy desk, table or other furniture. **HOLD** on to it and be prepared to move with it. Hold the position until the ground stops shaking and it is safe to move. Stay clear of windows, fireplaces, and heavy furniture or appliances. Don't rush outside. You may be injured by falling glass or building parts. DO **NOT** try using the stairs or elevators while the building is shaking or while there is danger of being hit by falling glass or debris.

2 If you are outside, get into the **OPEN**, away from buildings and power lines.

3 If you are driving – **STOP** if it is safe – but stay inside. DO **NOT** stop on or under a bridge, overpass or tunnel. Move your car as far out of the normal traffic pattern as possible. DO **NOT** stop under trees, lightposts, electrical power lines or signs.

4 If you are in a mountainous area, be alert for falling rock and other debris that could be loosened by the quake.

5 In a crowded public place, do **NOT** rush for the exits. Stay calm and encourage others to do so.

Even when there is no structural damage, the shaking that occurs during an earthquake can cause breakage and injury. Simple precautions include not hanging pictures, especially glass-fronted ones, over the head of one's bed, and securing china ornaments in place with 'earthquake wax'.

Prediction and prevention

The business of predicting earthquakes is not simple. Usually, the first rupture is the biggest, and the aftershocks in the following weeks are of decreasing magnitude. If, instead, earthquakes habitually built up towards a crescendo it would be more obvious when a big earthquake was due! One way to estimate *where* an earthquake is likely to occur is to map out areas where earthquakes have not been happening lately (Figure 3.4). If a length of fault can be identified with no recent break-points, but the lengths of fault to either side have many recent epicentres, then it is probable that a large amount of strain has built up in the seismically quiet bit in the middle. This strain will have to be released sooner or later, perhaps by a major rupture, like the snapping of an elastic band stretched to breaking point. The trouble is that the method of pinning down exactly *when* this will happen has so far proved elusive.

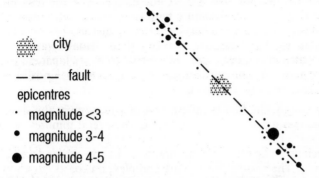

city

— — — fault

epicentres

·   magnitude <3

●   magnitude 3-4

●   magnitude 4-5

figure 3.4 Map of a fault and the surrounding area, showing earthquake epicentres over a 30-year period. The central part of the fault has not ruptured recently, so the prospects of a major earthquake epicentre in or near to the city in the next few years look pretty grim.

There is not much that can be done to prevent earthquakes happening. One suggestion has been to pump water into fault planes in areas where strain build-up is indicated by lack of recent seismic activity. The hope is that the water will act as a lubricant, allowing the crustal blocks to begin to slip past each other gently instead of violently. However, such a course of action is risky. Imagine the chances of re-election for a local government officer on whose orders this was done if, in order to forestall a magnitude 7 earthquake that was likely some time in the coming decade, the lubricated fault moved in such a fashion as to cause a series of magnitude 5 earthquakes. Even if only minor damage occurred, the body responsible for the fault lubrication programme would get the blame, and the lawyers would have a field day trying to establish financial liability!

Tsunamis

One other aspect of earthquakes deserves to be mentioned, and this is a potentially tragic consequence of earthquakes under the ocean. If the focus is at a shallow depth below the seafloor, the sudden displacement of rock can create a series of waves travelling through the ocean at a speed of several hundred kilometres per hour. In deep water, such waves may be scarcely noticeable, being less than a metre high and very gently sloped. However, upon reaching shallow, coastal water the waves slow down and become both higher and steeper with the result that they may spill over the shore with tremendous force, sweeping buildings (and people) away even tens of metres above sea level. Popularly (but incorrectly) called a 'tidal wave', this phenomenon is more correctly referred to by its Japanese name of tsunami. A tsunami can also be triggered by landslides or undersea volcanic eruptions.

Surprisingly, the first manifestation of a tsunami at the shore is sometimes a *fall* in sea level lasting for several minutes, which tragically can lure the curious onto the beach where they are especially vulnerable to the dramatic rise in water level that follows. The reasons for this are complex; on coasts close to the epicentre, the initial fall in local sea level may be because the land has been displaced upwards by the fault motion, but more generally it happens when the first part of a wave to arrive is a trough rather than a crest.

The wide-ranging effect of tsunamis is demonstrated by the case of the Hawaiian Islands, whose north shores face the earthquake-prone belt of Alaska and the Aleutian Islands, over 3000 km away. The largest historic tsunami to hit Hawaii happened in 1946, and was triggered by an earthquake in the Aleutians. This wave broke on the shore and reached up to 17 m above normal sea level, sweeping away entire villages and taking the lives of 159 people. Today, yellow tsunami-warning speakers mounted on telephone poles are a familiar sight in parts of the Hawaiian islands. It takes a tsunami several hours to travel from the Aleutians to Hawaii, but there is much less warning time in the case of tsunamis generated locally.

On 26 December 2004 a magnitude 9.1 earthquake, with its focus at a depth of 30 km in the Sunda Trench, offshore of Sumatra, happened when the rock either side of a 400 km length of fault jerked by about 10 metres, rupturing an area of fault surface almost the size of California. This triggered a tsunami that devastated many shorelines around the Indian Ocean, and took almost all of the 283 000 deaths attributed to this earthquake in Table 3.1. Tragically, there was no tsunami early warning system in the Indian Ocean. This could have saved tens of thousands of lives in Thailand, Sri Lanka and elsewhere. However, Sumatra and the nearby small islands, where more than 200 000 people were killed by inrushing water up to 30 m above the normal high-water mark, were so close to the source that even the best early warning system would have saved very few.

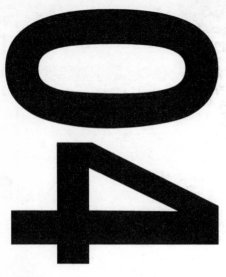

04

plate tectonics

In this chapter you will learn:
- how the Earth's outer shell is broken into plates that glide slowly across the weaker interior
- how this enables creation of new lithosphere and destruction of old lithosphere at plate boundaries marked by earthquakes and volcanoes.

We have so far discussed earthquakes without paying much attention to where in the world they tend to occur. Certain parts of the globe suffer more earthquakes than others, and you will have experienced at least a few minor tremors if you live in a high-risk area. A glance at a map of earthquake distribution, such as Figure 4.1, shows that they tend to occur in belts. The most famous of these stretches all the way up the west coast of the Americas, though the Aleutian and Kuril islands and southwards round the western side of the Pacific into New Zealand, so that the belt almost encircles the Pacific basin. This belt is also where many of the most active on-land volcanoes occur, and has been called the Pacific Ring of Fire.

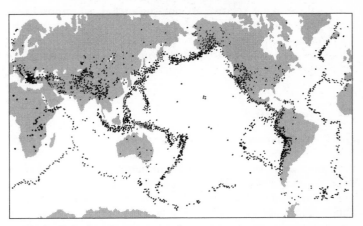

figure 4.1 Map showing the distribution of earthquake epicentres over a typical six-year period.

The coincidence of earthquakes and volcanic zones is no accident, and the relationship between these two quite different phenomena will become clear in this chapter and the next. Earthquakes also occur in a broad belt running through the Mediterranean and eastwards into central Asia (where there are very few volcanoes), and in some much narrower zones snaking their way through the oceans, notably a zone running from north to south through the Atlantic.

Moving plates

Earthquake zones are a manifestation of a phenomenon known as **plate tectonics**, though each of the three zones mentioned above fits into the picture in a different way. Plate tectonics is the term used to describe the manner in which the Earth's lithosphere moves around. As established in Chapter 02, the relative weakness below the uppermost mantle allows the lithosphere to slide around. However, the lithosphere is not a single unit. It is broken into a series of rigid plates, usually described as seven major plates and one group of six minor plates (these are named on Figure 4.9, on page 50). Most plates carry areas of both continental and oceanic crust. The Pacific plate is a notable exception in having no large areas of continent.

Each plate is in contact with its neighbours on all sides, but the plates are moving relative to one another. It is important to realize that there are no gaps between these plates, so there are no chasms open clear down to the asthenosphere. To see how plates are able to move around without any gaps appearing, we will look at processes at the margins of plates.

Plates colliding

First, what happens when plates are moving towards each other? This situation is typified by Japan, shown in cross section in Figure 4.2. Japan is a piece of continental crust on the eastern edge of a major plate. The floor of the Pacific Ocean belongs to a different plate, which is moving towards Japan at a rate of about 10 cm per year. Where the two plates meet, one is being thrust down below the other, a process described as **subduction**. Because oceanic crust is denser than continental crust, when two plates meet it is almost invariably the oceanic plate that goes

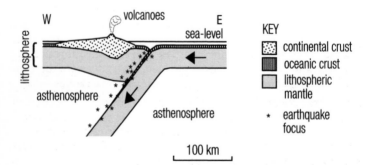

figure 4.2 Cross-section through Japan. See text for explanation.

under, as in this example. Under Japan, the Pacific Plate forms a slab descending at an angle of about 45°, but examples of much steeper and much shallower subduction are found in comparable situations elsewhere.

As one plate slides over another, the front of the over-riding plate is compressed, and the rocks there may become buckled. At the plane of movement between the two plates, motion is far from uniform, progressing in the stick–slip fashion that we considered in the previous chapter. This gives rise to earthquakes, and it was by plotting the depths of earthquake foci that subduction zones were first recognized for what they are. As you can see in Figure 4.2, earthquakes become deeper from east to west under Japan. A similar situation, though an east–west mirror image, occurs below Sumatra. Here the floor of the Indian Ocean is being subducted below Indonesia, and it was fault motion in that subduction zone (known as the Sunda trench) that led to the tragic earthquake and tsunami of 26 December 2004.

Inclined zones of earthquakes can be traced down to about 700 km, but no deeper. This is because, as a plate descends, heat from the surrounding asthenosphere warms it, so it eventually ceases to be recognizable. The first part of the subducting slab to lose its identity is its crust, because crustal rocks melt at a lower temperature than the temperature of the sub-lithospheric mantle that they encounter. The melted material rises upwards to feed volcanoes, which tend to occur in a belt about 70 km above the subduction zone. We will look at such volcanoes in more detail in the next chapter.

The top end of a subduction zone is marked by a trench on the sea floor. Off northern Japan the trench is a fairly typical 8 km deep, but further south the trench reaches over 11 km, which is the greatest known depth in the world's oceans.

When two plates meet at a subduction zone one of them is destroyed, so this setting is described as a **destructive plate boundary**. However, it is only the oceanic part of a plate that can be destroyed in this way. Let's consider what happens when both plates contain continents (Figure 4.3). At first, subduction proceeds as normal, while the oceanic part of one plate descends below the other. However, eventually the continental part of the subducting plate reaches the subduction zone. Continental crust is thicker and more buoyant than oceanic crust, and this prevents the plate from continuing to subduct. The edges of both continents are buckled, but one will eventually be thrust

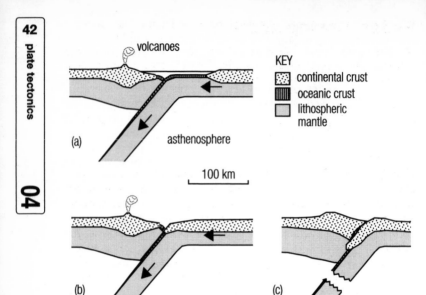

figure 4.3 Time series of cross-sections to show a continent–continent collision. See text for explanation.

over the other. Soon afterwards, the subduction zone jams. The oceanic part of the descending slab breaks free, leaving the two plates joined together above. Near this suture the crust may be double its normal thickness, and this is where the highest mountains are to be found.

A recent example of such a continent–continent collision began about 30 million years ago when India collided with Asia. We see the aftermath of this event in the high elevation of the Tibetan plateau, fringed to the south by the Himalayas. Another consequence of this collision, and others that accompanied it, is the diffuse belt of earthquakes running eastwards from the Mediterranean through central Asia. As for the vanished ocean, the only remains are a few slivers of oceanic crust and upper mantle caught up in the suture zone, known as **ophiolites**.

Plates moving apart

So much for what happens where two plates are moving towards each other. This cannot be the only sort of interaction between plates, because if it were the Earth's surface would have to be shrinking all the time. In fact, we now know that new ocean floor is being created at a rate sufficient to compensate for the loss of oceanic parts of plates at destructive plate boundaries. Unsurprisingly, these sites are known as **constructive plate boundaries**.

What goes on there is summarized in Figure 4.4. As two oceanic plates are drawn apart, in a process referred to as sea-floor spreading, the underlying asthenosphere wells upwards to avoid any gaps appearing. As the upwelling asthenospheric mantle approaches the surface it cools, and becomes part of the lithosphere belonging to the plates on either side of the boundary. This new lithosphere is still relatively warm, which makes it slightly less dense and more buoyant than the older, colder lithosphere further from the boundary, so these boundaries are marked by ridges. Typically, the crest of such a ridge lies at a depth of 2–3 km below sea level, whereas the expanse of ocean floor to either side is at an average depth of 4–5 km. There are earthquakes associated with the rifting and subsidence of oceanic lithosphere near constructive plate boundaries, but these are all at shallow depths, giving a seismic belt entirely different in character from a subduction zone.

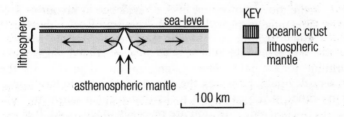

figure 4.4 Cross-section through a constructive plate boundary, where upwelling asthenosphere accretes to the diverging edges of two lithospheric plates. The oceanic crust is a result of partial melting of the upwelling asthenospheric mantle. See text for explanation.

In addition, a small percentage of the upwelling mantle melts. This is not because there is a heat source here, it is a consequence of the drop in pressure, and is known as **decompression melting**. When mantle of peridotite composition (about 45 per cent silica, SiO_2) begins to melt, the molten rock (or **magma**) produced has a slightly higher silica content. This is what gives rise to oceanic crust (Table 2.2) where the composition is, on average, about 49 per cent silica. This composition, described as 'basaltic', '**basic**' or 'mafic', results in the rock type known as **basalt**, and the residual mantle left behind has a reduced silica content to compensate for the enrichment of silica in the crust, but because the volume of melt produced is very much smaller than the volume of mantle contributing to the melt, the chemical change in the mantle is slight.

The ability of rock to melt in this way, so that a small volume of melt enriched in silica (and other components) is produced, is a very important process in geology, and is known as **partial melting**. The whole of the oceanic crust has been produced by partial melting of the mantle. We will look in more detail at how the oceanic crust is constructed, and in particular at the sub-sea volcanism associated with constructive plate boundaries, in the next chapter.

Magnetic stripes on the ocean floor

One of the most compelling lines of evidence in favour of sea-floor spreading is provided by magnetism. Rocks that have cooled down from a melt inherit their magnetization from the Earth's magnetic field as it was when they cooled. Such rocks forming today have their magnetization aligned with the present-day field. However, the direction of this field has flipped many times in the past, and rocks that formed when the field was in the opposite direction are reversely magnetized. It is not known why the field reverses, but it provides another argument, if one were needed, for discounting the idea of a giant bar magnet within the Earth.

Studies of the magnetization of the sea floor, carried on during the 1950s by towing sensitive instruments behind research vessels, began to reveal a curious stripy pattern of magnetization on the sea floor. Half the stripes are magnetized in the direction of the present field, but half are magnetized in the reverse direction. This was elegantly explained in 1963 by two British geoscientists, Fred Vine and Drummond Matthews, who suggested the explanation of a mid-ocean 'tape recorder' (Figure 4.5) which results in a 'bar code' pattern. As new ocean floor is added at a constructive plate boundary it is magnetized according to the direction of the Earth's magnetic field, and it retains this magnetization as it moves away. Reversely magnetized stripes represent ocean floor formed at times when the Earth's field was in the reverse direction.

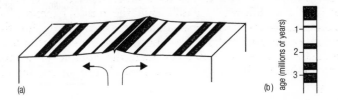

figure 4.5 (a) The mid-ocean 'bar code'. Normal (black) and reverse (white) magnetic stripes on the ocean floor show the polarity of the Earth's magnetic field at the time when each piece of ocean floor was added to the edges of the plates at a constructive plate boundary. (b) The magnetic polarity timescale.

The dates of reversals in the Earth's magnetic field extending back for a couple of hundred million years have been determined by studying the magnetization of volcanic rocks on land, whose age has been established radiometrically. As a result, the age of the ocean floor at any point can be found just by counting back the number of magnetic stripes from the constructive plate boundary, or by recognizing a distinctive 'bar-code' pattern of reversals of different lengths, without the need to collect a sample.

The break-up of a continent

If a linear belt of upwelling in the asthenosphere establishes itself below a continent instead of an ocean, the continental crust at first becomes stretched thin, then it ruptures, and eventually a constructive plate margin may develop in between, with formation of new oceanic crust. This is illustrated in Figure 4.6, and something of the sort seems to have been responsible for the origin of the Atlantic Ocean. The apparent fit of South America

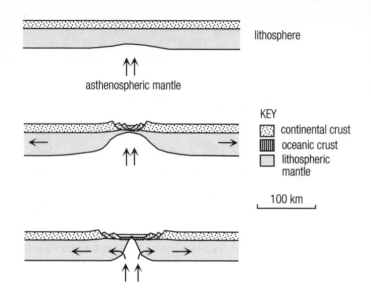

figure 4.6 Time series of cross-sections to show the splitting of a continent to form a new ocean. Once sea-floor spreading has begun, it continues at a rate typically between 1 cm and 20 cm per year. Subsidence along some of the faults that were initiated during the initial rifting event may continue sporadically at the rifted edges of the continents for tens of millions of years. This is the biggest cause of earthquakes not at plate boundaries.

and Africa if you imagine the Atlantic closed (and not just the fit of their coastlines, but of more fundamental geological structures too) has long been recognized. It gave rise to theories of moving continents, described as **continental drift** before the discovery of plate tectonics in the 1960s provided the explanation.

The life-cycle of an ocean

The rates of ocean floor creation, though varying from place to place, are always approximately equal on either side of a constructive plate boundary. In consequence, the ridge marking this line in the Atlantic Ocean is midway between the two shores. Constructive plate boundaries in general are therefore often called 'mid-ocean ridges'. This is a misnomer, because eventually the

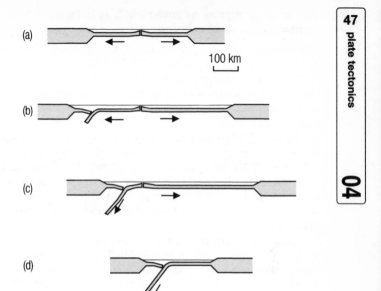

figure 4.7 Time series of cross-sections to show the evolution of an ocean, formed in the way shown in Figure 4.6. The scale is reduced here, and the crust has not been distinguished. The stage reached by the Atlantic Ocean is shown by (a), though it is wider than shown here. In stage (b) the left-hand plate has split, and the oceanic part is being subducted beneath the mainly continental part. The Pacific Ocean today is at stage (c).

oceanic part of one or other of the plates becomes detached from the continental part and begins to subduct beneath it (Figure 4.7), as has happened in the Pacific and Indian Oceans. Whether or not the ocean continues to become wider depends on the relative speeds of subduction and sea-floor spreading, but whatever happens the subduction process ensures that the constructive plate boundary does not remain central in the ocean.

Figure 4.7 shows what will almost certainly be the eventual fate of the Atlantic Ocean. If we imagine this as a west–east cross-section then it implies subduction (with its attendant earthquakes and volcanoes) beneath New York. However, it could equally well happen on the European side instead. Eventually, the constructive plate boundary could be drawn into the subduction zone and, once it has been removed in this way, the ocean must grow narrower until, as in Figure 4.3, there is a continent–continent collision.

It is thought that the cycle of continental splitting, ocean opening, ocean closing and continental collision takes on average 400–500 million years. You may have noticed when reading the above that whereas ocean floor is continually created and destroyed, there has been no mention of formation of *continental* crust. Continental crust is split and reassembled in ever-changing configurations by the action of plate tectonics. Although it can be worn away by the erosional part of the rock cycle (Figure 1.1) this leads to redistribution rather than destruction of material, so continental crust is never destroyed, and nor does it grow much these days except by the addition of small volumes resulting from volcanic activity above subduction zones. We will postpone consideration of the origin of continental crust until Chapter 13, except to say that most is thought to go back over at least 2 billion years, although much of it has been deformed many times since then, and that some traces date back nearly 4 billion years. In contrast, because it is recycled, the oldest known oceanic crust is a mere 190 million years old.

Plates sliding past each other

Not all boundaries between plates are sites of either convergence or separation. Some mark places where plates are sliding past each other. These are known as **conservative plate boundaries**. Figure 4.8 shows a typical example within an ocean, where a constructive plate boundary is offset by a fault, known as a **transform fault**. A transform fault may offset the ridge by anything from ten to a couple of hundred kilometres. It is marked by a narrow zone of shallow earthquakes. The transform fault is the bit between the two offset ridge segments; its apparent continuation beyond this lies within a single plate, and is not a plate boundary. This is known as a 'fracture zone' and is often manifested as a sharp change in depth because of the different ages of ocean floor on either side of it, but it has far fewer earthquakes because there is no sideways slip between the rock on either side.

Conservative plate boundaries can run through continents too. The most celebrated of these, of which the San Andreas fault is a part, runs through California. Here, the part of California on the south-west of the fault is attached to the Pacific Plate, and is

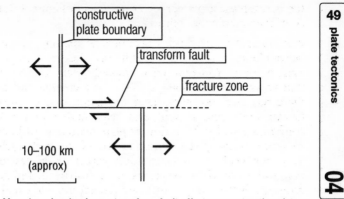

figure 4.8 Map view showing how a transform fault offsets a constructive plate boundary.

moving north-westwards at an average rate of 1 cm per year relative to the rest of North America. It is stick–slip motion along this fault system that causes most of the earthquakes in the San Francisco–Los Angeles region.

The global picture

We are now in a position to view the big picture of how the plates and the boundaries between them fit together on a global scale. This is shown in Figure 4.9. There are essentially two settings in which destructive plate boundaries can occur. The first is where one oceanic plate descends below another. When this happens, volcanism on the over-riding plate leads to the construction of a series of volcanoes, which is the origin of the volcanic **island arcs** in the western Pacific. On the other hand, when subduction occurs beneath a continent, as where the Nazca Plate (south-east Pacific) subducts beneath South America, the volcanoes grow on pre-existing continental crust, giving rise to an Andean-type volcanic mountain range. The case of Japan, with which we began, is something in between. Part of Japan is old continental material rifted away from the eastern edge of the Asian continent, but much of it is young island arc material.

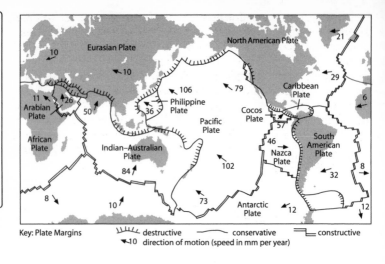

figure 4.9 Global map on the same base as Figure 4.1, showing plate boundaries and the rates of plate motion relative to Africa (which is virtually stationary). Recent continental collision zones are shown here as destructive plate boundaries.

A potentially confusing feature of any map such as Figure 4.9 is that it attempts to portray on a flat sheet of paper something that is really happening on the surface of a sphere. Tectonics occur on a sphere, which means that rates of relative plate motion must vary along a plate boundary. For example, the constructive plate boundary between the Pacific Plate and the Antarctic Plate is spreading at a rate of less than 6 cm a year between Antarctica and New Zealand, but the rate gradually doubles as the boundary is traced over the next 1000 km north-eastwards. The relationships between individual plates are continually adjusting to prevent any gaps opening between plates. This effect is particularly noticeable in trying to trace through time the positions of triple junctions where three plate boundaries meet.

So far we have kept the discussion simple, by assuming that the relative motions between plates are parallel. Although this is commonly a good approximation, especially for constructive plate boundaries, it does not always hold. There are plenty of examples where collision between continents occurs obliquely. Indeed, much of the western coast of North America from California to Alaska is thought to consist of pieces of continent described as 'exotic **terranes**' which slid in obliquely from the south at various times during the past 200 million years. To take another example, conservative plate boundaries within the continents often have kinks in them. In the example shown in Figure 4.10, the relative motion as seen looking from one side of the fault to the other is from left to right. Overall, the fault runs parallel with the direction of relative motion, but where the fault line bends to the left there is a zone of compression (manifested by buckling of rocks and thrusting of one unit over another) and where it bends to the right there is a zone of extension where the crust is stretched thin. The San Andreas fault system is of this type.

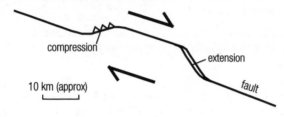

figure 4.10 Map view to show zones of local compression or extension where a conservative plate boundary bends one way or another.

What makes it happen?

In all this discussion of plate tectonics, we have not considered what drives it. This is still controversial, but most geologists would agree on two things. The first is that, attractive as the idea might seem, plate motions are not the direct surface expressions of mantle convection of the kind illustrated in Figure 2.9. The plates seem to be moving around more or less independently of this deeper process. The second is that, whatever the immediate driving mechanism, the creation of new, hot oceanic lithosphere at constructive plate boundaries and the subduction and recycling of old, cold oceanic lithosphere at destructive plate boundaries is the principal means by which the heat generated by radioactive decay in the mantle escapes to the surface. Probably about twice as much heat gets out this way as trickles out through the lithosphere by conduction, and each of these considerably outweighs the outward heat transfer associated with volcanic eruptions. One of the great unsolved mysteries of modern geology is why Earth behaves in this way but Venus, which is almost the same size, mass and density as the Earth, seems to lack plate tectonics.

There are many forces that could be causing plates to move. As already noted, simple conveyor-belt-type drag by flow in the underlying asthenosphere is not considered likely. Nor does it seem that plates are pushed apart by forcible injection of new material along constructive plate boundaries. It is probably closer to the truth to regard the upwelling here as a consequence rather than a cause of plate divergence. Perhaps the most likely driving mechanism is that the old, cold edge of a subducting slab, sinking because it is negatively buoyant, drags the rest of the plate with it.

We have mentioned volcanoes several times so far. In the next chapter, we will consider them more fully.

05

volcanoes

In this chapter you will learn:
- about the different kinds of volcanoes characteristic of various geological settings
- what makes them erupt, and what hazards are posed by different kinds of eruption.

We saw in the previous chapter that the world's most obvious volcanoes are above subduction zones in Andean-type mountain chains or island arcs. The global distribution of volcanoes above sea level (Figure 5.1) correlates well with destructive plate boundaries (Figure 4.9). There is rather more volcanic activity, though of a different kind and harder to detect because it is usually several kilometres deep underwater, occurring continually along constructive plate boundaries. In contrast, conservative plate boundaries rarely show associated volcanism. In this chapter we will consider the types of volcanic activity that happen at plate boundary settings, and also look at volcanism that occurs well away from the edges of plates.

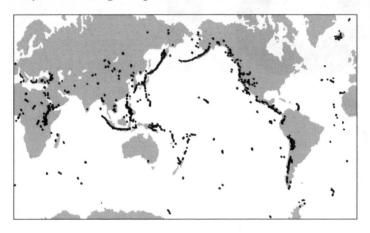

figure 5.1 Global map of volcanoes above sea level, showing volcanoes known to have erupted in the past 10 000 years.

Mention the word 'volcano', and most people probably think of something looking like the volcano in Figure 5.2; a fairly steep conical mountain like Mount Fuji in Japan. This sort of volcano, known as a composite cone volcano or stratocone, is typical of settings above subduction zones, though they are not all so symmetrical.

Magma origin and eruption

In subduction zone volcanism, much of the material that is being melted at depth belongs to the crust of the descending

figure 5.2 A composite cone volcano, Acamarachi in northern Chile. The date of its most recent eruption is not known, but is believed to have been within the past 10 000 years.

slab. With the exception of any sediments that have accumulated on top of it, this will be basaltic in composition, implying a silica content of about 49 per cent (Table 2.2). Most of the melt derived from this is richer in silica (about 55–60 per cent) as a result of partial melting, a process we met in the previous chapter. Melts of the **intermediate** silica content produced by partial melting of basalt are called **andesites** (named after the Andes mountains, where they are common).

As it begins to rise, this intermediate magma will often mingle with batches of basaltic magma generated by partial melting in the wedge-shaped zone of mantle between the top of the subducting plate and the base of the over-riding plate. Melting is stimulated here as a result of water having been squeezed out of the wet crust of the subducting plate. This water rises into the overlying mantle wedge, where it has the effect of lowering its melting temperature. This is described as **hydration melting**, and (like decompression melting) is a way of creating magma without actually heating the source material.

Magma is usually slightly less dense than solid rock, so once formed, it tends to rise towards the surface. At first it percolates along the interfaces between crystals, but eventually collects into larger blobs that can force their way upwards. Some magma bodies cool sufficiently to solidify before reaching the surface, forming intrusions, which we will look at in the next chapter. For now, we are concerned only with what happens if the magma reaches the surface (Plate 2). When it does, one of two things may happen. It may ooze out and begin to flow downhill. This is described as a 'lava flow', **lava** being the term used to describe magma once it has reached the surface. Alternatively it may fragment, sometimes explosively, to produce a **pyroclastic** (fire-broken) rock.

Magmatic gases and explosive eruptions

The idea of lava exploding may seem surprising. It is, however, a very common occurrence (Figure 5.3, Plate 3). What makes lava explode is the sudden escape of gases that had been dissolved in the magma. If these gases escape rapidly (like a can of fizzy drink opened after it has been shaken), bubbles form that expand the lava into a froth and may break it into

figure 5.3 An eruption from the south-east crater of Mount Etna, Sicily. Gases are escaping from magma within the vent with force sufficient to drive this mild pyroclastic eruption. Fine fragments are thrown to a height of about 50 metres above the rim of the crater, and then blown downwind in a billowing cloud.

fragments. Lumps of solidified bubbly rock are referred to as pumice. The small fragments are referred to as volcanic **ash**, which correctly describes their fine-grained nature and typical greyish colour but is potentially misleading because, unlike the ashes of a fire, volcanic ash is not a product of combustion.

Magma rising above subduction zones typically contains a few per cent of gas dissolved within it. Typical gas composition is about 56 per cent H_2O (water vapour), 28 per cent CO_2 (carbon dioxide), 14 per cent SO_2 (sulphur dioxide) and one per cent or less of each of H_2S (hydrogen sulphide) HCl (hydrogen chloride), H_2 (hydrogen) and CO (carbon monoxide). Above a subduction zone, most of the dissolved water vapour can be explained as sea water escaping after having been transported down with the descending slab (in fissures and hydrated minerals in the oceanic crust or within wet sediments), and some of the carbon dioxide probably comes from breakdown of subducted carbonates (which are common in certain kinds of sedimentary rock, most notably in limestone). However, magmas in other settings also contain dissolved gas (though usually less) and it seems that some of the gas is escaping from within the mantle. This represents the tail end of the degassing process that is thought to have been the source of most of the atmosphere (see Chapter 02).

Whether a volcanic eruption is explosive depends not on whether the magma is shaken like beer in a can, but on how quickly the magma rises and how easily the gases can escape. The great pressure at depth keeps gases dissolved in the magma when it is deep down, and usually bubbles begin to form only within a couple of kilometres of the surface. Events during the final ascent from this depth are crucial. If the magma rises slowly, then there will usually be time for it to degas quietly, and on reaching the surface the magma will form a lava flow. If it rises quickly, or if the walls of the conduit are impermeable to gas, the gas will remain trapped within the magma and it will erupt explosively.

In many volcanoes, gas escapes quietly over periods of tens or hundred of years without any associated eruption. The fissures where the gas reaches the surface are known as fumaroles, and may be at temperatures of up to several hundred degrees centigrade (Figure 5.4).

figure 5.4 Volcanologists studying fumaroles on the summit of Vulcano, a volcanic island to the north of Sicily that gave its name to volcanoes in general. On mixing with air, the volcanic water vapour forms a visible steam plume. The volcanologists are wearing gas masks to protect against additional acid gases, notably sulphur dioxide, in case the wind shifts.

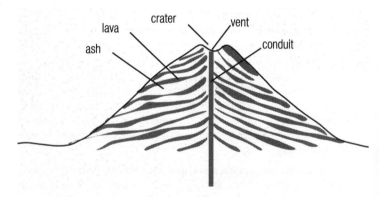

figure 5.5 A cross-section through a composite cone volcano, showing the alternation of ash and lava responsible for both its name and it shape. There is only a single vent shown here, but in many cases the conduit branches at depth, feeding secondary ('parastitic') vents on the flanks of the volcano.

Eruption conditions in andesite composite cone volcanoes tend to fluctuate between quietly effusive and explosive. The resulting alternation of lava flows and more widely dispersed ash layers is chiefly responsible for the relatively steep conical shape of this kind of volcano (Figure 5.5).

Volcanic hazards

Lava flows of andesite composition are fairly viscous. They tend to have very blocky surfaces, and move more like a train of rubble than a stream of liquid. As a result, they are rarely longer than about 10 km. Although they may damage property, people can usually escape because such flows normally advance at only a few metres per hour.

When andesitic volcanoes erupt explosively, the effects are much more widespread. One of the biggest eruptions of the last century occurred in June 1991 at Mt Pinatubo in the Philippines. This volcano had not erupted for at least 500 years. As is often the case, when it awoke it erupted much more dramatically than a similar volcano in the habit of erupting every few years, and a column of ash reached a height of about 60 km over the volcano. This type of eruption is described as plinian. The name comes from Pliny the Younger, a Roman youth who wrote the first stage-by-stage account we have of a

figure 5.6 A plinian eruption column rising about 20 km above Lascar volcano, in the Andes.

volcanic eruption, that of Vesuvius in 79 AD, which culminated in the destruction of the towns of Pompeii and Herculaneum (and caused the death of his uncle, Pliny the Elder). A plinian eruption column of an intermediate size is shown in Figure 5.6.

During a plinian eruption, the ash is driven upwards by the force of the explosive escape of gas near the vent. However, columns of ash would not attain the great heights that they do without the aid of buoyancy. What happens is that air is drawn into the column and heated by the hot ash particles and by mixing with the hot volcanic gases. The heated air expands and, despite the weight of the ash particles, the column as a whole becomes buoyant and rises until it reaches a height where its density matches that of the surrounding air. The eruption cloud is then blown downwind, and the ash begins to settle out to form an 'airfall' ash deposit. The larger particles fall faster, so airfall deposits tend to be thicker and made of larger particles closer to the volcano (Figure 5.7). The largest bits, ranging up to metres in size, are known as volcanic bombs. By mapping out the thickness of ancient airfall deposits it is possible to determine the wind direction at the time of the eruption and also to estimate the intensity and volume of the eruption. Where it accumulates thickly, airfall ash can cause roofs to collapse, and the fine particles can cause choking, but it is rarely a major cause of death.

figure 5.7 This volcanic ash seen in cross-section on the island of Vulcano, Italy, represents several eruptive pulses. In each case, the larger particles fell first, so that each layer becomes finer upwards.

Sometimes, an eruption column becomes unstable and all or part of it collapses onto the side of the volcano, and then sweeps downslope as a cloud of searing hot ash and gas travelling at over 100 km per hour. This sort of **pyroclastic flow** is known as a nuée ardente (French for 'glowing cloud') and can also be triggered by the collapse of a steep dome of extruded lava. A nuée ardente is justifiably the most feared consequence of a plinian eruption. One reason for this is the destruction of the town of St Pierre, the capital of the Caribbean island of Martinique, and the death of all but two of its 29 000 inhabitants, in the 1902 eruption of Mt Pelée. One survivor was a prisoner in a windowless jail cell, partly below ground level. He was dug out badly burned two days later, and pardoned for his crimes. Fears of a similar eruption on the nearby island of Montserrat led to the temporary evacuation of its most vulnerable areas in 1995, followed by a permanent evacuation in 1996 when the island's capital was abandoned (Figure 5.8).

figure 5.8 Progressive burial of Plymouth, the capital of Montserrat, mainly by pyroclastic flows. Top: April 1997. Lower left: July 1998. Lower right: May 1999.

In the event, the Montserrat dome collapse flows were smaller but more persistent than on Martinique. Many people left the island, and with the eruption still continuing, in 2006 the island's population was less than half what it had been ten years previously.

Unlike airfall deposits, which blanket the topography irrespective of slope, pyroclastic flows tend to be confined to valleys, a characteristic that enables the two types of deposit to be distinguished. However, not all airfall material remains in place. In particular it is easily washed away by rainfall. In tropical areas rain is liable to be torrential, so the water courses fill with a slurry of ash and water that, being denser than water, moves with a great force. These volcanic mudflows (often called lahars; an Indonesian word) caused most of the damage resulting from the 1991 eruption of Mt Pinatubo. Lahars can also be initiated if a volcanic eruption melts a glacier or snowcap. This is what happened during an eruption of Ruiz, Colombia, in 1985 when approximately 23 000 people,

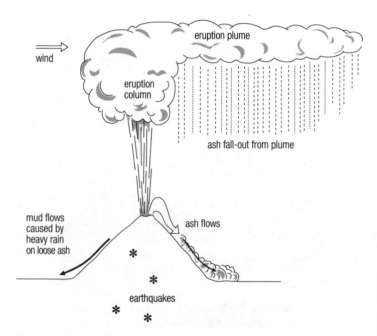

figure 5.9 Some of the hazards associated with explosive eruptions. The most dangerous ash flows are nuées ardentes.

including most of the population of the town of Armero, lost their lives. Readers who are old enough may remember harrowing television pictures of a young girl, her lower body trapped by solidified debris in the mudflow, and with only her head showing above water, who succumbed after several days' attempts to dig her out.

The principal hazards deriving from explosive eruptions of andesitic volcanoes are summarized in Figure 5.9. We may also add the risk of volcanic collapse, a process that first became widely recognized after the May 1980 eruption of Mt St Helens, in Washington State, USA. Here, two months' warning of a major eruption was given by small but progressively shallower earthquakes, minor ash eruptions and a slight bulging of the northern flank of the volcano. However, rather than culminating in an anticipated plinian eruption from the summit crater, the northern flank gave way, forming a giant debris avalanche. As soon as this collapse began, the confining pressure on the magma within the volcano was released and it degassed explosively, directing most of its force sideways. This directed blast was supersonic, and 60 square kilometres of what had been forest was devastated (Figure 5.10).

figure 5.10 Mt St Helens, three years after its 1981 eruption. The whole summit was undermined and destroyed by the collapse and directed blast. The steam within the collapse amphitheatre comes from a slow-growing lava dome that may eventually restore the volcano to its previous symmetrical cone shape.

The sideways nature of this directed blast caught even the professionals by surprise. A volcanologist working for the US Geological Survey, Dave Johnson, was stationed in what should have been a safe location, on a high ridge several kilometres north of Mt St Helens, having been assigned the duty of recording the eruption. Tragically, his campsite caught the full force of the directed blast, and his body was never found. His young field assistant, Harry Glicken, had been sent back to headquarters, and survived to become a well-known volcanologist in his own right. Sadly Harry died, 11 years later, when he was caught by a nuée ardente generated by a collapsing lava dome on the Japanese volcano Unzen.

The Mt St Helens collapse was triggered by the injection of magma into the volcano, but it is possible that other volcanoes collapse simply under their own weight. When this happens, debris avalanches can travel for tens of kilometres, to judge from mapping of ancient deposits. No major volcanic debris avalanche has happened in historic times.

Other causes of death associated with volcanoes include the release of suffocating gases, which may flow downhill, and tsunamis generated not by earthquakes but by underwater volcanic explosions, or by a volcanic landslide entering the sea or a lake. Table 5.1 lists the fatalities from some notable volcanic events.

Not all volcanoes above destructive plate boundaries are andesitic composite cones. In some places basaltic melts approach the surface, either because of more complete melting of the oceanic crust of the subducting slab, or because of partial melting within the wedge-shaped volume of mantle between the two plates. This basalt may flow out as quiet lava flows, its lower viscosity allowing it to spread more thinly than andesite over many square kilometres or, if the eruption is driven by the force of escaping gases, it may result in a 'scoria cone', formed of lumps of slightly frothy basalt (Figure 5.11).

On the other hand, very large volumes of more silica-rich magmas are sometimes generated above subduction zones. This is particularly likely where large proportions of sediment have been dragged down with the subducting slab, or when there is a lot of melting of the lower crust above the subduction zone. These magmas typically have around 70 per cent silica, and are described as **acidic** or granitic. These usually crystallize at depth, forming the well-known rock type called **granite** which we shall examine in the next chapter, but sometimes they do approach the surface.

Volcano	Year	Fatalities	Main cause(s) of death
Vesuvius, Italy	79AD	>3500	pyroclastic flows
Kelut, Indonesia	1586	10 000?	unknown
Asama, Japan	1598	800	religious pilgrims killed at summit
Vesuvius, Italy	1631	>4000	pyroclastic flows
Merapi, Indonesia	1672	3000?	pyroclastic flows
Laki, Iceland	1783	9350	starvation
Asama, Japan	1783	1500	pyroclastic flow, lahars
Unzen, Japan	1792	14 300	tsunami
Tambora, Indonesia	1815	92 000	starvation
Krakatau, Indonesia	1883	36 400	tsunami
Mt Pelée, Martinique	1902	29 000	pyroclastic flows
Taal, Philippines	1911	>1335	pyroclastic flows
Merapi, Indonesia	1930	1369	pyroclastic flows
Ruapehu, New Zealand	1953	151	lahar
Iliwerung, Indonesia	1979	539?	tsunami
Mt St Helens, USA	1980	57	directed blast, lahars
Mayon, Philippines	1981	>200	lahar
El Chichon, Mexico	1982	1900	pyroclastic flows
Ruiz, Colombia	1985	23 000	lahar
Lake Nyos, Cameroon	1986	>1700	asphyxiation by gases
Pinatubo, Philippines	1991	800	airfall, lahars, disease
Soufriere Hills, Montserrat	1997	19	pyroclastic flows
Casita, Nicaragua	1998	1600	lahar
Nyiragongo, Congo	2001	147	lava flows
Merapi, Indonesia	2006	2	pyroclastic flows

table 5.1 Some notable volcanic events.

figure 5.11 A basaltic scoria cone, in the Andes of northern Chile.

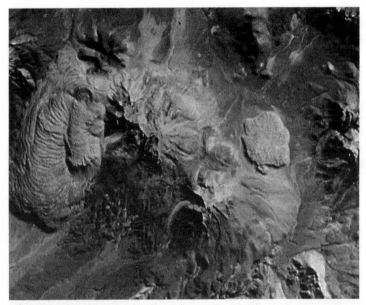

figure 5.12 An image of a 40-km-wide area in the Andes, recorded by satellite. On the left is a 24 cubic km acidic flow, whose front is 300 m high. Its thickness and the large wrinkles on its surface indicate that it was an extremely slow-moving viscous flow. On the right is an acidic dome, only one-tenth the volume, which appears to have oozed out radially from a hidden vent below its centre.

It is rare for granite magmas to ooze out quietly. This is because granitic lava is even more viscous than andesitic lava, making it difficult for gases to escape quietly. Sometimes it does happen though, and a very thick flow or a steep-sided lava dome is formed (Figure 5.12).

Ignimbrite eruptions

When granitic magma approaches within a few hundred metres of the surface, the gas pressure usually fractures the roof over the magma body, thereby releasing pressure and encouraging whatever gas remains in solution to come out all at once. This drives the most energetic variety of explosive eruption known, which occurs at volcanoes colloquially known as 'supervolcanoes'. Such eruptions are known from the study of ancient deposits rather than from observation because (fortunately for us) they are infrequent and no example has occurred in recent times. To judge from the airfall deposits that we find preserved, a high plinian eruption column develops, sometimes feeding a continent-wide ash cloud. The most characteristic deposit is probably a result of column collapse.

figure 5.13 The high cliffs in the background of this view expose a thick section through an ignimbrite erupted from the Valles caldera, New Mexico, USA. The lower cliff in the foreground is a plinian airfall deposit. Note the person for scale in the centre.

This takes the form of a very extensive sheet of acidic ash, in which the particles are often welded together because they were still very hot when emplaced. This sort of deposit is known as an **ignimbrite**. There are good recent examples in the south-west USA (Figure 5.13), New Zealand and the Andes, and 400-million-year-old examples in Snowdonia (Wales) and the Lake District of England.

Large ignimbrites are the result of the eruption of up to around 100 cubic km of magma. This tends to leave a large hole in the ground, rather than an obvious volcanic mountain. Usually what had been the roof over the magma body collapses to form the floor of a large crater, termed a **caldera** (Figure 5.14).

figure 5.14 Cross-section to show caldera formation. Between these two stages, a large-volume ignimbrite-forming eruption occurs.

I have described volcanism near destructive plate boundaries in broad outline. Unfortunately, we cannot explore the many details of eruptive processes in a book of this length. We turn now to volcanism at constructive plate boundaries.

Volcanoes associated with continental rifting

First, what happens when a continental plate is in the process of splitting to initiate a new constructive plate boundary? Figure 4.6 showed that this happens above a linear zone of upwelling in the asthenosphere. The faults that form and allow the crust to stretch are actually rather good pathways along which magma can rise upwards, and indeed continental rifting seems often to be preceded by an approximately 10-million-year period of eruption of mainly basaltic lavas, produced by partial melting of the upwelling mantle asthenosphere. This is going on today in the African Rift, a large fault-bounded structure that runs from Eritrea and Ethiopia southwards through Kenya and

Tanzania to Mozambique. Africa appears to be splitting apart by east–west extension across this rift, although there is no guarantee that this will progress far enough to generate a new ocean.

This extension provides the upward pathways for the magma feeding the large African volcanoes like Kilimanjaro. Asthenospheric upwelling is at its most intense near the northern end of the African Rift, corresponding to a **hot spot** over a plume rising from deep within the mantle, below the point where the rift joins the Red Sea and the Gulf of Aden. Here the eruptions have taken the form of vast sheets of basaltic lava flows, known as continental **flood basalts**.

There are many ancient examples of flood basalts that appear to represent hot spots prior to continental rifting. One of the largest and most famous is the 65-million-year-old Deccan Traps, in India. These are a kilometre thick. Although reduced by erosion, they still cover an area of half a million square kilometres, and are signs of a hot spot that helped India to rift away from Africa and Madagascar. (The hot spot is still active,

figure 5.15 The island of Staffa, Scotland, formed of a 60-m-thick basalt flow, the middle part of which (from present sea level up to the roof of the cave) fractured into columns as a result of contraction during comparatively slow and orderly cooling. The upper part is more rubbly. The cave is Fingal's Cave, made famous by Mendelssohn in his overture *The Hebrides*.

being marked by the site of the volcanically active island of Reunion.) Other continental flood basalt provinces reflect the rifting of the Atlantic Ocean, and can be found split into unequal parts on either side. These include: the 125-million-year-old Parana basalts of southern Brazil and their less-extensive counterpart, the Etendeka traps, in Namibia; and the smaller but more famous 60-million-year-old Brito-Arctic province, most of which is in east Greenland, but which includes the thick lava flows of the Giant's Causeway, in Northern Ireland, and the Inner Hebrides islands, Scotland (Figures 5.15 and 5.16).

figure 5.16 Top view of columnar joints formed by slow cooling within the flood basalt flow on Staffa, Scotland. Most columns are hexagonal, though five-side examples can also be seen.

The 15–17-million-year-old Columbia River Basalts of Washington, Oregon and Idaho constitute a continental flood basalt province above a hot spot that has not been followed by a major rifting event. One of these flows travelled 300 km along the Columbia River Gorge before spreading out laterally over 100 km.

Volcanism at ocean ridges

Once true ocean crust has begun to form, by sea-floor spreading, the character of the volcanism changes. The basaltic

figure 5.17 Pillow lava, erupted on the ocean floor 95 million years ago at a constructive plate boundary. This is now exposed on land, as part of an ophiolite in the Oman mountains, Arabia, which are a rare example of a slice of oceanic lithosphere that instead of being subducted was thrust on top of continental crust during a collision.

lava that erupts at ocean ridges to form the top layer of the oceanic crust can take on a form completely unlike that of lava on land. When lava is extruded under water, its surface chills very rapidly, forming a flexible rind around the still-molten interior. The lava typically resembles a pile of water-filled plastic bin-liners, a morphology known as '**pillow lava**' (Figure 5.17, Plate 4).

Where lava is erupted fast enough at a constructive plate margin, it may form sheet flows. Alternatively, it may fragment to form a glassy debris known as hyaloclastite. However, explosions, as opposed to fragmentation, are virtually unknown because the pressure exerted by the overlying depth of water inhibits the violent escape of gases.

One special characteristic of volcanism at constructive plate margins is the widespread occurrence of vents where superheated water emerges. This happens where water that has percolated down through the crust over a wide area is warmed and then expelled, by convection, through vents along the axis of the ridge. On the way it dissolves some elements from the rock, and certain other elements already dissolved in seawater come out of solution by reacting with the rock. The hot water emerges as an acidic solution, but as soon as this mixes with seawater some of its dissolved constituents become insoluble, and form a smoke-like cloud of metal sulphide particles. The most famous examples are vents at about 350 °C where the particles are black, hence their name 'black smokers'. These hot vents are surrounded by organisms like clams, crabs and 'tube-worms' that ultimately depend not on plants that derive their energy from sunlight, but on bacteria-like microbes that, in turn, feed by oxidizing the sulphide particles. It has been suggested that life on Earth (and elsewhere) could have originated in such a setting, and only later spread to sunlit regions where plants learned to photosynthesize (see Chapter 12).

We noted in the previous chapter that the partial melting at a constructive plate margin is caused by the decompression associated with the upwelling of asthenosphere at a rate necessary to replace the lithosphere drawn away by subduction. However, when a length of constructive plate boundary is situated above a plume coming up from deep within the mantle, the volume of melt produced is much greater than usual. As a result the oceanic crust may be two or three times thicker than normal, and its surface will be above sea level. The only really clear example of this today is Iceland.

Hawaiian volcanism

We turn now to volcanoes that are unrelated to plate boundaries, and consider the example of Hawaii. A plume from deep within the mantle is involved again, but this time the plume hits the base of the lithosphere more than 1000 km from the edge of the Pacific plate. This plate is moving north-westwards over the plume, and as a result there is a chain of increasingly older, extinct volcanoes stretching north-westwards from the Big Island of Hawaii where the current activity is centred. With time, the weight of these older volcanoes causes them to subside, and they are also worn down by erosion, so only the youngest few remain above sea level. The rest of the chain survives only as undersea peaks known as seamounts.

During the most productive part of its lifetime, a Hawaiian volcano erupts nothing but basalt. Basalt is the least viscous type of lava. There are two important consequences of this for a volcano above sea level. First, any gas within the lava can escape relatively easily, so eruptions are not usually highly explosive. Second, the lava flows, being so runny, are able to spread out thinly. They can also flow a long way before cooling down, especially if the top of the flow freezes and then acts as an insulating lid, allowing the molten lava to continue to flow in a 'lava tube' within. In consequence, Hawaiian volcanoes are characterized by gentle slopes of 2–3 °. Such an edifice is known as a **shield volcano**, because in profile it resembles a shield laid on the ground (Figure 5.18). When lava flows reach the coast their fronts become chilled by the seawater, so they tend to spread sideways along the shore. This steepens the shoreline to the extent that eventually it collapses, and much of the submarine portion of the cone is made of collapsed lava rubble rather than intact lava flows.

Basalt lavas like those erupted on Hawaii take essentially two forms. When moving relatively fast, because of a fast eruption rate or a locally steep slope, the surface breaks into clinkery blocks. This type is known by the Hawaiian term 'a'a' (pronounced 'ah-ah', and said by some to represent the cries of pain made by someone trying to walk barefoot over such a jagged surface). When moving more slowly, a smooth, chilled but pliable skin forms over the molten interior of the flow, and the lava spreads as a series of smooth lobes described by another Hawaiian word, pahoehoe (Plate 5), which translates as something like 'flat swirly swirly'. This does not fragment like a'a, although drag by flowing lava beneath the skin can pleat the surface over upon itself to form a pattern reminiscent of coiled rope. Both a'a and pahoehoe are visible in Figure 5.19.

50 km

figure 5.18 The profile of a Hawaiian-type shield volcano. Mauna Loa volcano on the Big Island of Hawaii counts as the highest volcano on Earth, because it rises more than 9 km from its base on the floor of the Pacific Ocean.

figure 5.19 Two kinds of basalt lava on Hawaii: clinkery a'a (left) and ropey pahoehoe (right).

Lava hazards

Basaltic lava flows, even though they are the least viscous and therefore fastest flowing of common lava types, usually advance at less than five metres per second. They rarely kill people because there is usually plenty of time to get out of the way. Tragic exceptions have occurred when pent-up basalt lava has escaped rapidly from breeches in the flank of Nyiragongo volcano in the Congo, taking a estimated 60–300 lives in 1977 and 147 in 2002. Normally, though, it is property and livelihoods rather than lives that are threatened by basaltic volcanism. Lava flows are virtually unstoppable, and easily overcome barriers erected to try to pond or divert them. One of the more notable successes in lava flow diversion was in 1992 when the front of a major lava flow on Etna, Sicily, had reached within a few hundred metres of the town of Zafferana. The side of the flow was deliberately breached by explosives part way along its length. This caused the flow to spread out sideways high up on the mountain and halted its downstream advance.

Predicting eruptions

About 60 volcanoes erupt each year. A tenth of these eruptions are over within a day and some go on for years, but the average duration is about seven weeks. How can we tell when an eruption is due, and how can we predict what its effects will be?

The best way to anticipate an eruption's effects is to study previous eruptions. This is easy in the case of volcanoes that erupt every few years, because there are likely to be eyewitness reports, video recordings and so on. However, it is the volcanoes that erupt least frequently whose eruptions are usually the most dangerous, because a greater volume of ash and/or lava tends to come out in one go. In such cases, there may be no historical records and the volcanologist must depend largely on the traditional skills of the field geologist – specifically, interpretation of past events based on detailed mapping and logging of ancient deposits. In addition, computer programs exist that can use a digital model of the terrain to predict the paths of ash, lava and mud flows. Hazard zone maps based on these approaches are available for many volcanoes.

There are many precursor signs of an impending eruption. If recognized, these can give days' or even months' warning. In this respect volcanoes are easier to predict than tectonic earthquakes in which, as we have seen, the first shock is usually the biggest. Probably the single most useful eruption precursors are special kinds of small earth tremors, readily distinguished from tectonic earthquakes, whose depth typically decreases from several kilometres to a few hundred metres during the build-up to an eruption. These are associated with injection of magma into the volcano, or with migration of gas bubbles. In order to pinpoint them it is necessary to have an array of at least three seismometers on or close to the volcano.

Seismometers are expensive instruments, and few volcanoes have permanent arrays. At the time of the first historic eruption of Pinatubo, Philippines, in 1991 the volcano had been little studied and was not under surveillance until a small ash eruption near the summit in March showed that it was waking up. An array of portable seismometers was then rushed into place, enabling the locations of sub-volcanic tremors to be tracked as they approached the surface. This gave enough notice for the most vulnerable areas to be evacuated before the catastrophic eruption in June. Most of the ensuing deaths were attributable to the failure to sweep airfall ash off flat roofs (which collapsed under the weight) and subsequent mudflows, and not due to lack of eruption warning.

When magma is forced into a volcano, the volcano swells in size. In extreme cases its contours may change by many metres. This deformation is another useful precursor sign, and its pattern may serve to indicate whether the eruption is likely to occur at the summit or from a vent or fissure on one side. It can be studied by ground-based surveying or by using radar data from satellites.

Other techniques include measuring tiny changes in gravity that reflect injection or withdrawal of magma from a volcano, measuring changes in the rate of gas emission or of gas composition, and monitoring changes in surface temperature using ground-based instruments or satellite-borne infrared detectors.

Widespread effects of eruptions

Volcanic eruptions don't just affect people living locally. Several intercontinental airliners have run into trouble by unwittingly flying into volcanic ash clouds, whereupon their engines have stalled. Luckily, on every occasion so far the pilots have been able to restart the engines, though sometimes only after a terrifying gliding descent most of the way to the ground. Eruption cloud warning systems, based on images from weather satellites, and specially designed aircraft-mounted radar are now used to protect the most vulnerable air traffic routes, notably the transpolar route between Europe and Japan that passes over Alaska and the Kurils, and routes to Australia via Indonesian airspace.

The biggest volcanic eruptions can have continental or even global effects. The 1991 eruption of Pinatubo injected so much fine ash and sulphur dioxide into the stratosphere that global surface temperatures are estimated to have been reduced by about half a degree centigrade throughout the following year. This is because sunlight that would normally have warmed the surface was blocked in the upper atmosphere. The largest eruption in modern history was of Tambora, Indonesia, in 1815. At least 50 cubic km of ash was erupted, which was responsible for glorious red sunsets round the globe and a much greater drop in temperature than after the Pinatubo eruption. The following year, 1816, has been described as 'the year without a summer' by northern hemisphere commentators. There were widespread crop failures and probably more deaths than from any other historic eruption.

When we consider the likelihood of a similar eruption in the near future, not to mention an even larger eruption of a supervolcano, we have to accept that volcanoes could play a big role in short-term climate change. The possibility of worldwide famine cannot be discounted.

Volcanic activity also has more benign effects. Magma at shallow depths can heat the ground water and cause it to gush episodically through a vent in the form of a geyser, like the archetypal Geysir in Iceland or Old Faithful in Yellowstone Park, Wyoming. Volcanically heated water can also be piped to provide free heating to nearby communities, as is done in Iceland and parts of Japan. Heated water may also seep out gently to form hot mud pools. Bathing in these is said to be therapeutic, though personally I have found the clinging smell of rotten eggs (caused by hydrogen sulphide) to be an annoyance that seems impossible to get rid of, no matter how long I shower for afterwards!

Hot springs are as much a manifestation of magma intruded at depth as of magma erupting at the surface, and we will look at these intrusions in the next chapter.

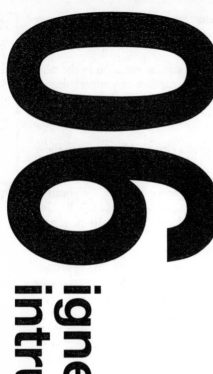

06

igneous intrusions

In this chapter you will learn:
- some basic information about the minerals of which rocks are composed
- about crystallization of minerals as magma cools slowly underground
- how this affects the properties of the remaining magma and the kinds of rocks that this produces.

Rocks formed by cooling of magma are called igneous rocks, from the Latin word for fire. We were dealing with igneous rocks of the volcanic kind throughout the previous chapter. However, not all magma reaches the surface. Often it solidifies at depth, to form an **intrusive** igneous rock.

Volcanic rocks cool so quickly that often the only crystals that grow within them are too small for the unaided eye to see. However, an intrusive rock cools more slowly, so there is time for the crystals of which the rock is formed to grow to lengths of millimetres or centimetres before the rock is completely solidified. Crystals this big are easy to see when the body is eventually exposed on the surface as a result of erosion. It is time now to consider the crystalline nature of igneous rocks, before looking at the sorts of intrusive bodies they can form.

Minerals

Because the crust and mantle contain so much silicon and oxygen (Table 2.2), all the common sorts of crystals that grow in the usual magmas from these sources contain these elements. The crystalline structure of rock-forming igneous **minerals** is based on two- or three-dimensional arrays of silicon and oxygen held together by molecular bonds. These minerals, as well as the rocks they form, are commonly referred to as silicates.

The most familiar of the silicate minerals is pure silica. This contains just silicon and oxygen, and has the formula SiO_2. In crystalline form this is the mineral quartz, except at extremely high pressures when the atoms are squeezed into more tightly packed crystalline structures. All other silicate minerals contain various metallic elements in addition to silicon and oxygen. Quartz is a robust, hard-wearing mineral whose significance you will realize in Chapter 07. It is able to grow in magmas whose SiO_2 content exceeds about 52 per cent. Quartz is abundant in rocks of granitic composition and is usually present, though scarce, in rocks of andesitic composition.

The potentially confusing use of silica to mean two quite different things is standard in geology. The silica (or SiO_2) content quoted for a rock is based on an overall chemical analysis of the elements making up the rock. This should not be taken to imply that all or any of the silicon and oxygen occurs in the form of quartz or any other pure silica mineral. On the other hand, when SiO_2 is given as the chemical formula of a mineral, it means that this mineral contains Si and O only, in the ratio 1:2. This specific ratio is dictated by the number of oxygen atoms with which each silicon atom is able to form a bond. The three-dimensional structure of such bonds controls the properties of the crystal.

In magmas whose composition is basic (45–52 per cent silica) or **ultrabasic** (<45 per cent silica), metallic elements are so abundant that no quartz crystals can grow, and the only silicate minerals to form contain a relatively high proportion of metals. The most extreme example is olivine, formula Mg_2SiO_4, in which there are twice as many metal atoms (in this case, magnesium) as silicon atoms. Actually, iron (Fe) atoms can fit in the same place as magnesium atoms, so olivine can have any composition ranging between Mg_2SiO_4 and Fe_2SiO_4, with Mg and Fe in any proportion. The permitted variation in olivine's formula is often expressed by quoting it as $(Fe,Mg)_2SiO_4$. This degree of variability in composition, which is achieved without changing the internal structure of the crystal, is what distinguishes a mineral from a chemical compound. Usually in igneous rocks olivine is of a magnesium-rich variety, say $Fe_{0.2}Mg_{1.8}SiO_4$.

Olivine most often occurs in association with pyroxene, another metal-rich silicate, with the formula $(Ca,Mg,Fe)_2Si_2O_6$. A rock composed essentially of just olivine and pyroxene is described as a peridotite, and these are the dominant minerals in mantle rocks.

In magmas of higher SiO_2 content, the proportions of magnesium and iron tend to be lower, whereas aluminium (Al), calcium (Ca), sodium (Na) and potassium (K) become more abundant. One result is the appearance of minerals known as feldspars. Plagioclase feldspar, the most abundant mineral in rocks of andesitic composition (52–66 per cent SiO_2), has a formula ranging between $NaAlSi_3O_8$ and $CaAl_2Si_2O_8$ (a continuous range in composition is made possible by 'double substitution', whereby Na and Si are replaced by Ca and Al in equal proportions). Potassium feldspar ($KAlSi_3O_8$) occurs chiefly in granitic rocks (>66 per cent SiO_2), in association with quartz and plagioclase feldspar.

The other important silicate minerals are mica and amphibole. These occur in granitic and andesitic rocks and can form only if there was water present in the magma. They have complicated formulae: muscovite mica is $KAl_2(AlSi_3O_{10})(OH)_2$, biotite mica is $K(Mg,Fe)_3AlSi_3O_{10}(OH)_2$, and amphibole is $(Na,K)Ca_2(Mg,Fe,Al)_5(Al,Si)_8O_{22}(OH)_2$.

Each of these types of mineral, and the varieties within each type, can be distinguished by features such as colour, hardness, **cleavage** (regular planes of weakness in the crystal resulting from its atomic structure) and optical properties. A few characteristics of the main minerals are summarized in Appendix 1. With practice it is possible to recognize many of these when you find a coarse-grained igneous rock exposed in the field, without the aid of analytical instruments. Having identified each of the minerals making up a rock, it is then possible to give the rock itself a name, based on the criteria summarized in Appendix 2, and thence to deduce the conditions under which the rock is likely to have formed.

We will say no more about the mineral building blocks of igneous rocks here. The collection and study of minerals is an enthralling pastime in its own right, and there are many books and websites dealing with the subject (see Taking it Further, p272).

Crystallization of minerals in cooling magma

As a magma rises it gets cooler, because heat escapes into the surrounding rocks. Eventually the temperature and pressure will be appropriate for minerals to start crystallizing. The nature of these minerals depends on the overall composition of the magma. The growth of mineral crystals in a magma is not like the freezing of water, where the whole liquid solidifies at a specific temperature. Instead, each mineral species begins to crystallize at a different temperature. Generally speaking, the more metal-rich minerals crystallize at higher temperatures, and quartz and the sodium- and potassium-rich feldspars crystallize at lower temperatures. This process, which is effectively partial melting in reverse, is known as **fractional crystallization**.

If the crystals are carried along with the remaining magma until everything has crystallized, then the overall composition of the eventual rock is the same as that of the initial magma, even though the rock consists of several minerals each having a different composition. However, if the first crystals become separated from the magma (perhaps because they settle to the bottom, or because the magma is squeezed out), the remaining magma will have a different composition, richer in silica than that it started with. In this way andesitic magmas can evolve from originally basaltic magmas (Figure 6.1), and granitic magmas can evolve from andesitic magmas. When these new magmas crystallize, they will form rocks different in composition from the initial magma.

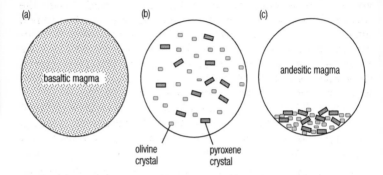

figure 6.1 Schematic cross-section (not to scale) showing how a body of basaltic magma (a) can begin to grow crystals of olivine and pyroxene (b). If these crystals begin to settle out, as in (c), they from a rock of ultrabasic composition and the remaining magma (which may crystallize later) is andesitic ('intermediate') in composition.

At the surface, a basaltic magma begins to crystallize at about 1200 °C and is entirely solidified by about 1080 °C. The corresponding temperatures for a granitic magma are 1100 °C and 960 °C, and those for an andesitic magma lie somewhere in between. At depth, the pressure acts to increase these temperatures by about 2 °C for every 10 km depth. However, this applies only for water-free magmas. Above destructive plate boundaries, escaping subducted water means that magmas are usually rich in water. The effect of this is to reduce the temperature at which crystallization begins, so that in a 'wet' granitic magma at 10–30 km depth, crystals do not begin to form until the temperature has dropped to about 720 °C. On approaching the surface, the water vapour is liable to escape, as we saw in the previous chapter. If the magma had already cooled to close to its 'wet' crystallization temperature, then this loss of water will cause the magma to solidify rapidly, because it will be well below its 'dry' crystallization temperature.

Granite

Crystallization promoted by volatile loss is one reason why granitic magmas in particular are more widely known for the large intrusions they form than for volcanic equivalents. Another reason is that they are about a thousand times more viscous than basalts, and so are less easily extruded onto the surface. It is not uncommon to find coarse-grained solidified intrusions of granitic composition that are tens of kilometres across. These constitute the well-known rock type granite (Plate 6). In fact, the name granite is so well known that monumental stone masons and the like tend to use it for any crystalline rock, whether or not its mineralogical composition fits the geological definition of granite. Poets use the term even more loosely, to refer to any rock that they consider to be hard and enduring, although (as we will see in Chapter 08) granite is not particularly immune to the effects of wind and rain.

How then does a granitic magma originating, say, 30 km deep near the base of the crust, rise upwards? As magma forms it seeps along grain boundaries, but cannot flow freely enough to escape until about five per cent of the source rock has melted, which may take 10 000 years or more. If there is a nearby plane of weakness, such as a fault or fracture, the magma may escape up it in a matter of 1000 years or so, and spread out higher in the crust to form a large granite mass. However, if there is no easy pathway available, magma will remain trapped at depth until it has grown into a large enough body to force its way up to a shallower level by buoyancy alone. This is thought usually to require melting of about a third of the source rock. At first, the surrounding rocks will be very hot, and soft enough to allow the granite magma to push them aside, rising at a rate of a few centimetres or metres per year. As the granite reaches progressively shallower depths, the surrounding rocks will be colder and therefore more brittle so that eventually the granite can rise no further.

There are other processes that may aid magma ascent. First, if the overlying rocks have a melting point similar to or lower than that of the magma, they may become assimilated into the magma. Essentially, the magma melts its way through, continuing to rise until it has run out of heat. This, incidentally, is another way in which magma composition can evolve. Alternatively, the granite may pluck away blocks of rocks from above. These sink down through the granite, allowing the granite to pass upwards. This process is called 'stoping', and is shown in Figures 6.2 and 6.3. The stoped blocks, which can be of any size, may either sink to the bottom of the magma unscathed, or may be melted by the heat of the magma and become incorporated within it. A third mechanism is apparent in the Himalayas, where granites are squeezed upwards along gently inclined faults related to the collision between India and Asia.

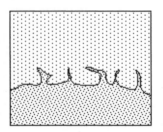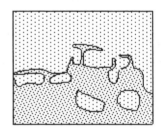

figure 6.2 Sketch cross-section to illustrate stoping at successive intervals. A rising body of magma plucks blocks off the roof to make way for itself. The scale could be anything from a metre to several hundred metres across.

figure 6.3 A pale granitic rock that became solidified in the act of stoping its way upwards into a darker rock type. The geological hammer gives the scale.

Most of the granites familiar to geologists ceased to rise at depths of a few kilometres within the crust. They have become exposed at the surface, millions of years after their emplacement, when the original overburden or rock has been worn away. This is encouraged by the fact that, even when solid, granite is less dense than most other rock types, and it tends to buoy up the terrain isostatically, increasing its vulnerability to the processes of erosion.

Granites can form wherever the composition of originally basaltic or andesitic magmas has been able to evolve to become sufficiently rich in silica, and so they can be found in any setting. However, they are particularly common above destructive plate margins (caused in particular by the melting of the lower crust above a subduction zone) and in the mountain belts at sites of continental collision. Here, the thickness of the continental crust is usually doubled (Figure 4.3), and because uranium, thorium and potassium are concentrated in the crust, the rate of local radiogenic heat production is doubled too. This encourages melting within the thickened crust even after relative motion between the two collided plates has ceased.

Large granite massifs such as that shown in Figure 6.4 are termed 'plutons', after Pluto, the Greek and Roman god of the underworld. Sometimes, geophysical evidence suggests that

figure 6.4 Half Dome, a famous granite pluton exposed by erosion in Yosemite National Park, California. The sheer face, beloved of climbers, is a result of erosion taking advantage of cooling joints.

several plutons are joined together at depth, in which case the larger unit is referred to as a 'batholith'. The relationship between plutons and a batholith is shown in Figure 6.5.

Not all plutons are granitic in composition. For example, the 60-million-year-old continental flood basalts that we met in the previous chapter (reflecting the rifting of the north Atlantic Ocean between Britain and Greenland) are associated with plutons of basaltic composition. Being course grained, we describe them as gabbro rather than basalt (see Appendix 2).

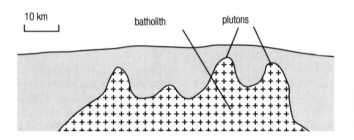

figure 6.5 Cross-section into the crust, showing several plutons as upward projections from a more extensive batholith at depth.

Basaltic intrusions in the oceanic crust

Gabbro is also crystallizing to form the lower oceanic crust at constructive plate boundaries. This gabbro layer can be recognized seismically in the oceans and can be studied on land in ophiolites, which, you may remember from Chapter 04, are slices of oceanic crust and upper mantle that have been thrust over the edge of a continent during a collision. It was once thought that the magma chambers within which these oceanic gabbros crystallize are ten or more kilometres wide. However, studies in the ocean attempting to map seismic reflections from the roof of these chambers, or to locate the molten volumes where S-waves cannot penetrate, have demonstrated that these magma chambers tend to be narrow, possibly ephemeral, features. Much of the gabbro appears to arrive in place as a crystal-rich mush.

Between the gabbro layer and the basaltic pillow lavas which erupted on the sea floor, there is a layer that is neither plutonic nor volcanic. This marks the depth at which pulses of magma are injected upwards along a vertical fissure to feed the lavas. When magma freezes in such a fissure it forms a vertical curtain of intrusive igneous rock called a **dyke**. In sea-floor spreading, successive dykes intrude either beside or within each other, and so a layer is formed consisting of nothing but dykes, which is sometimes called a sheeted dyke complex. The relationship between these layers in the oceanic crust is shown in Figure 6.6.

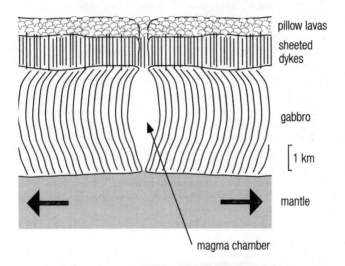

pillow lavas

sheeted dykes

gabbro

1 km

mantle

magma chamber

figure 6.6 The constitution of the oceanic crust, showing a cross-section through a spreading axis. Crust on either side is moving apart. Basaltic magma cools slowly at depth to form the coarse-grained basaltic rock type known as gabbro. The curved lines represent successive walls of the magma chamber, which can maintain a constant volume only if the rate of magma supply from below keeps pace with the rate of sea-floor spreading. Some of the magma is injected up fissures to the sea floor where it is erupted to form basalt lava (often taking the form of pillow lava), having the same composition as the gabbro, but a much finer grain size because it cools more rapidly. Magma that solidifies within the fissures forms vertical curtains of medium-grained rock, about a metre wide and called dykes.

Dykes and sills

Dykes cool faster than plutons, because they are thin curtains and can lose their heat to either side more easily. As a result, the crystals grow smaller than in a plutonic rock but coarser than in a volcanic rock. Dykes intruding other rock types are common in volcanic areas of the continents, but sheeted dykes, where dykes intrude nothing but dykes, can form only where there has been continual spreading, and are known only in ophiolites and below the ocean floor.

Because they are injected up cracks or fissures, dykes record tension in the crust at the time of their formation. In many volcanic provinces dykes occur roughly parallel to one another (at right angles to the direction of crustal extension), forming a dyke swarm. Near a main volcano it is common for dykes to radiate away from the volcano. Volcanic eruptions that occur on the flank (side) of a volcano rather than at the summit vent are usually where the top of a radial dyke reaches the surface. Another form of dyke occurs where magma is injected up the fracture that has allowed the floor of a caldera to subside (Figure 5.14). This is known as a ring dyke, because of its circular plan view. Not all shallow intrusions cut up vertically like dykes. Sometimes magma is injected in a horizontal sheet, especially where the surrounding rock is composed of flat-lying beds. The generally horizontal igneous body that results is described as a **sill**. The relationship between dykes and sills is shown in Figure 6.7.

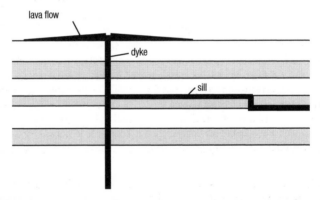

lava flow

dyke

sill

figure 6.7 Schematic cross-section showing the difference between a dyke and a sill. Both are sheet-like, tabular bodies, but a dyke is vertical and tends to cut across any bedding in the rocks it intrudes (described as a discordant intrusion). In contrast, a sill is intruded parallel to bedding (concordant intrusion), though it may have the occasional discordant step, as in this example.

Grain sizes in igneous rocks

We have made a generalization in this chapter that plutonic rocks are coarse-grained, shallow intrusions are medium-grained and volcanic rocks are fine grained. Although invariably true for plutonic rocks, there are circumstances where these rules are broken for shallow intrusions and volcanic rocks. For example, if magma begins to crystallize by slow cooling at depth but is then erupted, it may carry with it some largish crystals that had already begun to grow. These will be surrounded by much finer crystals that did not begin to grow until the magma cooled rapidly at or near the surface. Recognition of isolated large crystals (known as **phenocrysts**) in a lava or shallow intrusive rock provides evidence that the magma had a complicated history, cooling in at least two stages.

Another exception involves the very last fraction of a per cent of magma to crystallize in a magma body. This is likely to be very rich in water and volatile gases such as fluorine, because these tend to be excluded from crystals and so become progressively concentrated in the remaining melt as crystals grow. Strong concentrations of these make it difficult for new crystals to begin to form. As a result, the crystals that do get started have to grow very large, in extreme cases a metre or more long, but more usually centimetres in size. Such very coarse-grained igneous rocks are described as pegmatites, and can be found in **veins** (filling irregular fractures) within igneous intrusions or the surrounding rocks.

Conversely, not all volcanic rocks are crystalline at all. Magma erupted at the surface may chill so rapidly that there is no time for crystals to begin to form. The rock is therefore a frozen mixture of the elements that were present in the magma, lacking crystalline structure, and is described as **glass**. Pyroclastic igneous rocks may consist largely of glassy fragments, and many lavas consist of fine-grained crystals floating within a glassy matrix.

Shallow intrusions, especially dykes, although generally of medium grain size (with or without a few larger phenocrysts) often have very fine crystals, or even a glassy texture, at their edges. Such a **chilled margin** forms because the surrounding rock was cold, causing the edge of the igneous body to freeze quickly against it. Chilled margins are rare at the edges of plutonic intrusions. There are two factors contributing to this. The first is that, being deeper down, things are warmer in general anyway, so the surrounding rocks are not usually cold. The second is that plutons are so big and contain so much heat that they heat up the surrounding rock before being cooled much themselves. This heat can cause the surrounding rock to undergo dramatic changes through a process known as metamorphism, which is the subject of the next chapter.

07

metamorphism

In this chapter you will learn:
- how heat and pressure can cause dramatic changes to rock
- how these changes can be recognized, and what they can tell us about the processes that caused them.

All rocks are composed of minerals. These may be crystals that grew while a magma was cooling, as in the igneous rocks we considered in the previous chapters, or grains laid down to form a sedimentary rock by processes that we will look at in the next two chapters. Each mineral is stable only over a particular range of temperature and pressure. Beyond that range it will tend to break down or combine with neighbouring minerals to form new minerals.

Most igneous minerals are unstable at the low temperatures prevailing at the surface, and they tend to rot. This chemical **weathering** process is slow because, like most chemical changes, the speed of the reaction is slow at low temperatures. However, when a rock is held at a high temperature or pressure, its constituent minerals can change dramatically and thus completely alter the character of the rock. Recrystallization of this sort under the influence of temperature and/or pressure, but without melting, is known as **metamorphism**.

Thermal metamorphism

We closed the previous chapter with the example of a large igneous body intruding into colder rocks. When the rocks surrounding such an intrusion are examined in the field, they usually show the effects of heating by the intrusion across a zone extending a few hundred metres from the contact. This is described as contact metamorphism or **thermal metamorphism**.

Let's take the example of a granite intruding a fine-grained muddy sedimentary rock. Right against the contact with the granite we are likely to find that this mudrock has become very hard and splintery, and has clots of new minerals that grew within it as a result of contact metamorphism. A splintery, spotty rock such as this is described as a **hornfels**. Prominent among the metamorphic minerals is likely to be sillimanite, having the formula Al_2SiO_5. It grows only by metamorphism in rocks that contain aluminium and silica (a condition generally met in rocks formed from mud). As we walk away from the edge of the granite, we will eventually notice that the mudrock becomes less splintery, and the spotty pattern is lost. Soon the sillimanite disappears, and instead there is a different metamorphic mineral occurring as isolated stubby crystals up to a centimetre long. These crystals are of the mineral andalusite, whose formula is also Al_2SiO_5, but which grows at lower temperatures than sillimanite and has a different crystal structure. As we continue away from the intrusion, the andalusite crystals become fewer and smaller, until we find ourselves in unmetamorphosed mudrock. We have now walked right through the 'metamorphic aureole' of the granite.

Such a metamorphic aureole is sketched in Figure 7.1. Note, however, that sillimanite and andalusite form only in rocks of the right chemical make up. In rocks of other compositions, the metamorphic minerals will be different. For example, when a granite intrudes an impure **limestone**, the most prominent metamorphic mineral close to the contact is likely to be the iron-rich variety of olivine Fe_2SiO_4 (maybe $Fe_{1.8}Mg_{0.2}SiO_4$), which is unknown in igneous rocks.

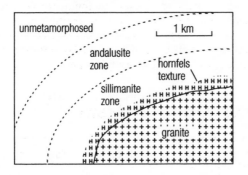

figure 7.1 Sketch map of the metamorphic aureole of a granite that was intruded into mudrocks. In this example, the outer metamorphic zone is characterized by the metamorphic mineral andalusite, which in the next zone inwards is displaced by sillimanite. Adjacent to the granite, the surrounding rocks have become brittle and developed a spotty hornfels texture.

Going back to metamorphism of rocks that were originally muddy, there is a third aluminium silicate mineral, known as kyanite, that can form if the pressure is high enough. The minerals sillimanite, andalusite and kyanite are described as **polymorphs** of the compound Al_2SiO_5; their crystalline forms differ because the atoms are arranged differently within them. Figure 7.2 shows the conditions of pressure and temperature under which each will form. From this we can see that the absence of kyanite in the metamorphic aureole of the granite in Figure 7.1 means that the pressure at the time of metamorphism must have been less than about 3 kilobars, equivalent to a depth of 10 km or less. Conversely, the aureole of a granite that was intruded deeper than about 15 km would have kyanite but no andalusite zone. In addition there might be high-pressure minerals of other compositions such as garnet, formula $(Fe,Mg)_3Al_2(SiO_2)_3$, which is sometimes of sufficient size and quality to be regarded as a semi-precious gemstone.

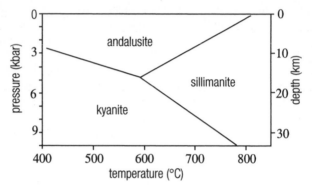

figure 7.2 The stability fields of the metamorphic minerals kyanite, sillimanite and andalusite. These all have the formula Al_2SiO_5, but are polymorphs stable over different ranges of temperature and pressure. The pressure scale is labelled in kilobars (1 kbar is equal to a thousand times atmospheric pressure), and the scale on the right gives the approximate conversion to depth in the crust.

You may wonder how it is that we can still find sillimanite surviving in the aureole of a granite exposed at the surface today, where clearly neither the pressure nor temperature are high enough for conditions to lie within sillimanite's stability field. The answer is that rocks tend to retain the minerals that grew within them at or near the conditions of highest temperature and pressure encountered during metamorphism. While a metamorphosed rock is cooling down there is usually insufficient impetus to drive the retrograde reaction from a high-temperature (or high-pressure) mineral to its low-temperature (or low-pressure) equivalent. This is fortunate, otherwise the only minerals we would ever see at the surface would be those stable at pressure of 1 atmosphere and about 0–30°C!

Regional metamorphism

Very often pressure, rather than temperature, is the main driving force behind metamorphism. Usually the two go together, because as depth (and hence pressure) increases, so does temperature. However, it is useful to distinguish thermal metamorphism, which is caused by proximity to a hot intrusion, from **regional metamorphism**, which is caused by the pressures and temperatures prevailing regionally. Any rock that is formed near the surface, whether volcanically or by deposition as a sediment, is liable to be buried by subsequent deposits. Eventually it may find itself at a depth where pressure and temperature are sufficient for metamorphism to begin. The whole of the lower crust consists of regionally metamorphosed rocks, except where it is made of recently intruded igneous intrusions.

Important departures from average conditions occur in collision zones (Figure 7.3). Where sediments have been dragged down into a subduction zone they will experience particularly high pressures, but having been recently at the surface they will not be as hot as most rocks at the same depth. They will therefore experience high-pressure, low-temperature metamorphism, and will be characterized by appropriate assemblages of minerals. On the other hand, where the crust is heated on a regional scale by large numbers of igneous intrusions, such as below the volcanic region near a subduction zone, crustal temperatures

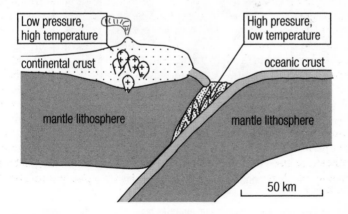

figure 7.3 Cross-section showing zones of low-pressure, high-temperature metamorphism, and of high-pressure, low-temperature metamorphism associated with a destructive plate boundary.

will be higher than normal at shallow depths. The rocks here experience low-pressure, high-temperature metamorphism. When a paired metamorphic belt such as this is found in ancient rocks, it is a good sign that there was once a destructive plate boundary there.

Metamorphic facies

Just as in thermal metamorphism, the metamorphic minerals that grow during regional metamorphism depend on the original composition of the rock. For example, a metamorphic mineral that requires magnesium cannot develop in a rock that lacks this element. Because of this, rocks that have been metamorphosed under identical conditions can contain entirely different minerals. Geologists therefore classify metamorphic rocks by groupings called **facies**. Each facies reflects a particular range of temperatures and pressures, but the minerals that develop in each facies differ according to the original rock type. The pressure and temperature conditions of the main metamorphic facies are shown in Figure 7.4. Metamorphism in

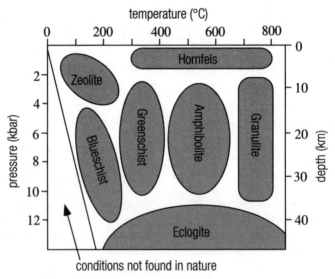

figure 7.4 The pressures and temperatures required to produce metamorphic rocks of each facies. These facies grade into one another, and the boundaries between them are ill defined, but they have been separated here for clarity.

the conditions near the top left of this diagram is described as low grade, and metamorphic grade is said to increase with distance from this point. Table 7.1 contrasts the most abundant minerals found in metamorphosed mudrocks with metamorphosed basaltic rocks in some of the facies named in this figure.

Facies	Originally mudrock	Originally basaltic rock
greenschist	muscovite, chlorite, quartz, sodium-rich plagioclase	albite, epidote, chlorite
amphibolite	muscovite, biotite, garnet, quartz, plagioclase	amphibole, plagioclase
granulite	garnet, sillimanite, plagioclase, quartz	calcium-rich pyroxene, calcium-rich plagioclase
eclogite	garnet, sodium-rich pyroxene, quartz	sodium-rich pyroxene, garnet

table 7.1 The mineral assemblages characterizing some metamorphic facies in two different pre-metamorphic rock types.

One thing that Figure 7.4 does not show is that, in the presence of water, granite crust will begin to melt at temperatures lower than those required for granulite facies metamorphism. When granulites are found, they are a sure sign of dry conditions in the lower crust during metamorphism.

Textures of metamorphic rocks

There is an alternative way to describe metamorphic rocks, which, for the geologist in the field, is simpler and more useful than facies. This is based simply on the texture of the rock, texture being a term that encompasses the size of the crystals, their shapes and their relative orientations. We have already met hornfels, which is a textural description for a fine-grained metamorphic rock resulting from contact metamorphism, in which spotty agglomerations of metamorphic minerals have grown.

In regional metamorphism, the rocks are affected by pressure as well as temperature. When this happens, crystals tend to line up so they lie at right angles to the direction of maximum compression. This is brought about by rotation of existing crystals, and growth of new crystals in this pressure-controlled orientation. This is most noticeable for those minerals that tend to grow as flat or elongated crystals, and less apparent in minerals whose crystals are more equidimensional. The alignment of minerals into layers is described as **foliation**, and such rocks are said to be foliated.

The lowest grade of regionally metamorphosed rock is **slate**, which is formed by zeolite facies metamorphism of mudrocks or fine silts. Slate is well known for its ability to be split thinly along flat foliation surfaces, hence its traditional use as a roofing material. These planes of weakness, or **cleavages**, running through slate are parallel to the metamorphic foliation. They have no relationship with any sedimentary layering that may have been present in the rock before it was metamorphosed. Slate is produced only from rocks that were originally fine grained and sedimentary in nature; if a basalt or a sandstone were to be metamorphosed in the same facies as a slate it would show few obvious signs of metamorphism, except maybe under microscopic examination. In particular, a coarse-grained rock cannot develop a closely spaced slatey cleavage.

If a slate is subjected to slightly higher grades of metamorphism, still mostly in the zeolite facies, but at higher temperature or pressure, fine flakes of metamorphic mica and chlorite minerals grow on the cleavage planes. These give the rock a greenish sheen. Typically the cleavages become wrinkled, and the rock would now be described as a phyllite.

At still higher grades of metamorphism, the metamorphic minerals grow bigger, and can be identified with the naked eye. Note the different significance of crystal size in igneous and metamorphic rocks. In igneous rocks larger crystals usually reflect slower cooling, whereas in metamorphic rocks larger crystals are a result of higher temperatures and/or pressures of metamorphism.

The next grade of foliated metamorphic rock after phyllite is called **schist**. A schist has metamorphic minerals readily visible to the naked eye. The foliation, which is wavy rather than flat, is often made conspicuous by concentrations of mica, which grows as shiny platy crystals. Schists produced by metamorphism under normal continental conditions, where temperatures tend to rise by about 30°C for each kilometre increase in depth, are formed under conditions of greenschist facies metamorphism, so named because of colour imparted by the green metamorphic minerals chlorite and (for rocks of basaltic origin) epidote. However, under high-pressure, low-temperature conditions that define the blueschist facies, a blue amphibole mineral (glaucophane) may develop. A schist texture can develop irrespective of whether the rock was originally fine grained or coarse grained.

Rocks metamorphosed under eclogite, amphibolite or granulite facies conditions are too far gone to contain abundant mica, and they rarely contain any other strongly planar minerals. In these the foliation is picked out by dark and pale minerals segregating into layers a few millimetres thick. This type of metamorphic rock is described by the German term '**gneiss**' (pronounced 'nice').

The most extreme metamorphic rocks are gneisses that reached the verge of melting. In these, veins and blobs of material crystallized from a melt are interspersed through the rock, usually strongly deformed and flattened into the plane of foliation. Such a rock is referred to as a migmatite.

The metamorphic textures described above can develop only in rocks that originally consisted of appropriate mixtures of minerals. There are two common rock types that are each made of essentially a single mineral, and so do not show these textures under whatever facies conditions they are metamorphosed. One of these is sandstone consisting of quartz grains. When a pure quartz sandstone is metamorphosed, the grains tend to fuse together producing a harder rock known as quartzite, but no new minerals can be formed. The other is limestone, which is a non-silicate rock consisting of calcium carbonate. When metamorphosed, limestone turns into marble, which can be a beautiful decorative stone. Recrystallized calcium carbonate usually gives it a bright white colour, and any impurities tend to concentrate into colourful veins.

Metamorphic rocks and most igneous rocks are products of processes within the crust. These are the hard rocks that often form the most prominent landscape features. Now we will begin to consider what happens to these rocks when they are exposed at the surface, and see how detritus derived from them accumulates to form the third main class of rocks, the sediments.

08

weathering, erosion and transport

In this chapter you will learn:
- how rocks are worn away by chemical and mechanical processes
- about the landscapes produced by erosion
- how rock fragments can be transported by wind, water and ice.

You saw in the previous chapter that each mineral is chemically stable over only a restricted range of temperature and pressure. Most minerals in igneous and metamorphic rocks form at temperatures and pressures well in excess of those at the surface, so they are no longer stable by the time the rock in which they are embedded becomes exposed. However, these minerals do not vanish in a flash (otherwise we would not be able to study them!). Their conversion to new minerals that are stable at low temperatures and pressures happens very slowly indeed. In dry conditions a mineral may survive hundreds of millions of years at the surface without appreciable change. The presence of water allows many minerals to 'weather' away rather more quickly, although the process may be virtually imperceptible on human timescales.

Chemical breakdown of minerals during weathering

Water controls the types of minerals that develop during weathering. Large crystals of igneous or metamorphic silicate minerals tend to rot away to tiny flakes of clay minerals. The clay family of minerals is very large, but clays all have characteristics in common. They are all hydrated, having the group OH, derived from water, in their formula, and often water itself, H_2O. Their crystal structure consists of sheets of silicate loosely bound together, which is what gives clay particles their microscopic flakey nature and makes a handful of damp clay so pliable.

The rotting of feldspar to produce a clay mineral may be described by the following chemical reaction between feldspar and water that is slightly acidic because of atmospheric carbon dioxide dissolved in it:

$$2KAlSi_3O_8 + 2CO_2 + 3H_2O = Al_2Si_2O_5(OH)_4 + 4SiO_2 + 2K^+ + 2HCO_3^-$$

feldspar	carbon dioxide (in solution)	water	kaolinite (a clay mineral)	dissolved silica	dissolved potassium	dissolved bicarbonate ion

This sort of reaction is described as **hydrolysis**. The clay (in this case kaolinite) remains as a solid residue, containing all the aluminium and some of the silicon that was in the feldspar, but the potassium and some of the silicon (as dissolved SiO_2) is washed away in solution.

Hydrolysis can take place both at the surface and at shallow depths where water is circulating through the rock. When carried to its extreme, and with similar reactions affecting other minerals, this can turn a granite into a deposit of kaolinite, which can be quarried as a source of china clay.

Hydrolysis contributes to the breakdown of micas, amphiboles and many metamorphic minerals. Potassium, sodium, calcium and magnesium tend to be removed in solution, but some can be retained in many clay minerals, like illite $KAl_3Si_3O_{10}(OH)_2$ or montmorillonite, which has the frighteningly complicated formula $(Na,Ca,Al,Fe,Mg)_8(Si_3O_{10})_3(OH)_{10}.12H_2O$.

When pyroxenes and olivines react with water, their magnesium and silica are carried away in solution, but their iron remains as a solid residue in the form of iron oxide such as the mineral hematite (Fe_2O_3).

The only common material that is more or less immune to chemical attack is quartz, which is pure SiO_2. Although SiO_2 is soluble in acidic groundwater, this is significant only for SiO_2 released by chemical breakdown of complex silicate materials during hydrolysis. Quartz itself is virtually insoluble under surface conditions.

Dissolved calcium and the bicarbonate ion resulting from hydrolysis may react and combine in a new environment to produce calcium carbonate, from which limestone is made. We will look at this in the next chapter. Other dissolved elements, notably magnesium, find their way into seawater and become incorporated into minerals in the oceanic crust by means of chemical exchanges that occur below hot vents like the black smokers we met in Chapter 05. There is thus a chemical strand to the rock cycle, involving solution and re-precipitation of elements. Although we have considered this in only the barest outline, we must now turn to physical aspects of the rock cycle.

Physical weathering

Rocks at the surface suffer not only the indignity of chemical attack, they also are exposed to the abrasive or grinding action of wind, water and ice. These are very effective agents of physical weathering. For example, if water repeatedly freezes and thaws in cracks and cavities, the expansion that occurs when water turns to ice can prise away fragments of rock. Even in dry conditions, a rock face may flake away because of cracking caused by expansion and contraction of the rock itself as its surface heats by day and cools by night. It may also be plucked away by the roots of plants. Fragments of rock embedded in a moving ice sheet (a glacier), or carried along by flowing water or wind, will abrade any surface they hit and, in turn, be abraded themselves.

Where there is a steep rock face, such as a cliff undermined by pounding waves, lumps of rock may fall and smash under gravity. Rocks that occur in weakly bonded layers, like slate that breaks easily along its metamorphic cleavage, or sedimentary rocks that were deposited in successive beds, are particularly prone to this – as are rocks in which closely spaced **joint** patterns have developed as a result of cooling (e.g. Figures 5.15 and 5.16) or release of pressure. However, the large blocks that accumulate in this way may be too big to be moved, even by strong waves or a fast-flowing river.

For transport to occur, rocks must usually first be broken into fragments smaller than a few centimetres across. Glaciers and landslides are notable exceptions, being capable of moving much larger pieces. On the other hand, the wind rarely blows hard enough to move anything coarser than a large sand grain.

When a fine-grained rock has been broken into transportable fragments, the particles may consist of several minerals still stuck together. Effectively they are just small pieces of rock. However, a transportable fragment of a coarse-grained rock is likely to be comparable in size with the crystals within it, so in what way does a coarse-grained rock break into fragments? You might think that each mineral would become separated from its neighbours. This can happen, but a characteristic of igneous and metamorphic rocks is that the minerals are usually intergrown and interlocking, like the pieces of a three-dimensional jigsaw, so they are really rather strongly held together. Such rocks usually break across crystals, rather than between them. What tends to happen in physical and chemical weathering is that pre-existing fractures within each crystal are attacked and opened. These fractures take two forms.

First there are cracks that may have been caused by deformation of the rock, or that opened up as the confining pressure grew less when the rock was brought closer to the surface. This sort of fracture may be found in any mineral. A more important sort of fracture occurs in minerals whose crystals contain inherent planes of weakness, called cleavage planes, caused by weak bonds in its crystalline structure. This is different from cleavage in a metamorphic rock, which is planes of weakness caused by alignment of platy minerals.

Common minerals with well-developed cleavages include the micas, feldspars, amphiboles and pyroxenes. In a mica crystal, all the cleavages are parallel, so it will tend to break into thin flakes bounded by cleavage surfaces. The others have cleavages in two or three orientations, and break into chunkier fragments.

Although larger rock fragments are bounded by fractures that cut across crystals, most sand-sized particles produced by the breakdown of a coarse-grained rock inevitably tend to consist of a piece of a single mineral. The smaller the particle, the greater its surface area compared with its volume, and so the more vulnerable it is to chemical weathering and the shorter its survival in the transport regime before it has rotted away.

We noted earlier that quartz is effectively insoluble and immune to chemical weathering. Another peculiarity of quartz is that it is the only common mineral to lack a well-developed cleavage. Quartz is therefore physically very robust, and it survives transport very well. It should be no surprise to you that sand grains on a river bed usually turn out to be mostly quartz, unless collected from close to their source, in which case there may be a fair proportion of feldspar and rock fragments too. The other solid residue (mostly clay particles) is unlikely to be found in the same place as this material, because being much finer grained it is washed away to settle out eventually in much quieter conditions, as you will see in the next chapter.

Sand-sized grains in wind and water

For now we will content ourselves with considering what happens to a sand-sized grain during transport (Figure 8.1). If broken off an igneous or metamorphic rock, a grain is likely to be quite jagged with sharp corners. This shape is described as angular. Each time a quartz grain bounces into a stationary rock, or collides with another grain, there is an opportunity for bits of the grain to be knocked off. Naturally, the corners are the most vulnerable parts. These get worn away first, so that the grain takes on a progressively more rounded shape. In contrast,

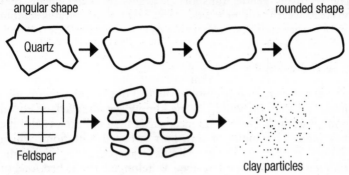

angular shape   rounded shape

Quartz

Feldspar

clay particles

figure 8.1 A grain of quartz starts off angular in shape, but becomes progressively rounded because of collisions during prolonged transport, which abrade its corners. On the other hand, a grain of feldspar is liable to fragment along cleavage planes, and the resulting small pieces will weather chemically, leaving clay particles as the only solid residue.

a feldspar grain is more likely to break into small fragments controlled by its cleavage planes, and hydrolysis will cause these to rot away to clay particles.

Currents of water and wind transport grains in similar ways. However, water is denser and more viscous than air, so it can move a particular sized grain in a current flowing much more slowly than the speed of wind required to move the same grain. For example, water flowing at only 0.2 metres per second (less than 1 km per hour) is capable of picking up 1 mm diameter sand grains, whereas wind needs to be blowing at 10 metres per second to do the same. This is 36 km per hour, and corresponds to wind force 5–6 (a fresh to strong breeze) on the Beaufort Scale.

In consequence, the collisions suffered by grains transported by wind are more violent than for water-transported grains. Sand grains blown around in a desert therefore tend to become very well rounded. Once there are no more corners to knock off, the effect of each collision is to make a microscopic pit on the surface at the point of impact, whose combined effect is to give the grain a frosted appearance. Seen in close-up under a microscope it looks just like etched glass. Another result of the high wind speed needed to transport sand grains is that rocks exposed to this 'sandblasting' may be sculpted into strange shapes.

The relatively gentle transport of grains by water means that physical weathering proceeds more slowly there. Grains seldom become very well rounded, their surfaces do not become frosted, and obstacles do not suffer from the sandblasting effect.

What wind and water have in common is the way in which they transport grains. Grains that are only just small enough to be moved will tend to stay close to the bed of the stream (in water) or the ground (in wind). Some simply get dragged or rolled along, but most of them bounce, suffering a collision each time they land. This mode of transport is called 'saltation', which literally means 'jumping'. Grains transported by dragging, rolling or saltation are described as belonging to the **bed-load** of the current. Smaller grains, once picked up by the flow, can remain suspended for long periods, suffering few collisions, and are said to belong to the **suspension-load**. An example might be a fast river in which sand grains are bouncing along its bed, constituting the bed-load, whereas fine clay particles are carried along in suspension, discolouring the water and forming the suspension-load.

Moving ice

Ice is a rather different transport medium. A glacier is ice flowing downhill at typical rates of between 20 and 200 metres per year, though occasionally a glacier can surge forward at over 300 metres a day. It is a powerful agent of erosion, plucking rock from the sides and floor of the confining valley. Glaciers also carry with them debris that has fallen from above. Once encapsulated in the ice, these fragments cannot bounce along, no matter how big they are. Ice-transported fragments thus tend to remain angular. The only way in which they can suffer abrasion is when they are carried at the base of the flow, in which case they will be scraped against the underlying rock. Debris transported by ice can often be recognized by scratches (or 'striations') on its surface, and corresponding glacial striations can be found on exposures of bedrock that once lay below a glacier.

We have now covered the effects of erosion and transport on the material that is being transported. We will conclude our review of this part of the rock cycle by looking briefly at its effects on the landscape.

Glacial landscapes

Staying with glaciers, we have already noted their ability to pluck away at the sides and floor of the valley down which they flow. A valley that has been sculpted by a glacier can be recognized by its U-shaped cross-sectional profile, very different from the more V-shaped valleys carved by rivers. Another peculiarity of glaciers is that they can scour deep hollows on the valley floor, in a way quite impossible for liquid water to achieve. After a glacier has melted, these hollows become filled with water, making lakes. Figure 8.2 illustrates the effects of glaciation on the landscape.

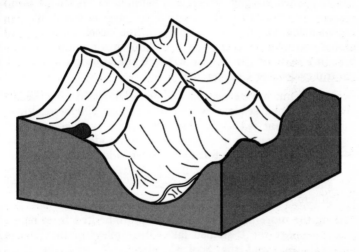

figure 8.2 A landscape that has been sculpted by the action of glaciers. The large U-shaped valley was carved by a major glacier, which filled the valley to a depth of some hundreds of metres. Tributary glaciers fed into it from the 'hanging valleys' on the side, whose floors are way above the floor of the main valley. The hanging valley on the left contains a mountain lake or tarn, which has filled a glacially scoured hollow. The ridges between adjacent hanging valleys are often quite sharp.

Glaciers, and more extensive ice sheets, today cover most of Antarctica and Greenland, with just an occasional ice-free peak sticking up above the level of the ice. Valley glaciers can be found in most of the world's high mountain ranges, and their rate of 'retreat' (because melting at the front outpaces resupply by ice-flow from the source) is one of the key indicators of global warming. However, there are large regions of the globe that are already ice-free where the landscape bears the unmistakable signs of glacial erosion. The areas affected include North America down as far as slightly south of the Great Lakes, and Europe as far south as southern Britain. This landscape was developed during the past three million years, when the polar ice sheets repeatedly advanced Equator-ward and then retreated.

Periods of extensive ice cover are referred to as **glaciations**, and the warmer intervals between are called **interglacials**. Repeating glaciations and interglacials make up an **ice age**, which is a climate phenomenon usually lasting about 10 million years that has happened roughly every 200 million years. There is no reason to think that the current ice age is over. We are experiencing an interglacial, though the accelerated rate of global warming in recent years probably has more to do with human release of greenhouse gases to the atmosphere than to natural cycles (see Chapter 13).

Because so much water was locked up as land ice during the last glaciation, global sea level was then more than 100 metres lower than the present sea level. Many glacial valleys were cut well below the present sea level, and are now flooded because most of that ice has melted to form long, steep-sided inlets of the sea, known by their Norwegian name of 'fiords'. Sea-level rise today has more to do with thermal expansion of seawater (as the global temperature creeps up) than with melting ice.

Among the more picturesque landscapes inherited from recent glaciations are the U-shaped lake-filled valleys of the English Lake District and the Scottish Highlands, and the fiord coastlines of Scandinavia, British Columbia and New Zealand's South Island. Among the bleaker heritage of those times are the relatively flat, glacially-scoured plains of central Canada and the 'channelled scablands' of Washington and Idaho (north-west USA), carved by repeated catastrophic floods when the alternately retreating and advancing ice sheet released the pent-up waters of glacially dammed lakes such as the 7000 square km Lake Missoula in Montana.

Erosion by rivers and streams

Beyond the sculpting of individual rocks and boulders that we have already referred to, wind is not often a sufficiently powerful erosive agent to have much effect on the landscape. In ice-free regions, it is liquid water that is the key player. Even regions with exceptionally arid climates suffer the occasional downpour. When this happens, valleys that may have been dry for a century or more become occupied by a torrential stream that may last only for a day or so.

Unlike glaciers, rivers are nowhere near deep enough to fill their valleys. Therefore they erode more at the base than the sides, which explains the V-shaped cross-sectional profile of a typical river valley (Figure 8.3). Rivers tend to flow less straight than glaciers, so when a valley is carved by a river it is likely to be sinuous, with spurs of the hills on either side overlapping. This means you cannot often stand on the floor of a river valley and see straight along it as you can in a glacial valley. If a glacier forms in a valley previously cut by a river, it will tend to straighten the valley by truncating the overlapping spurs.

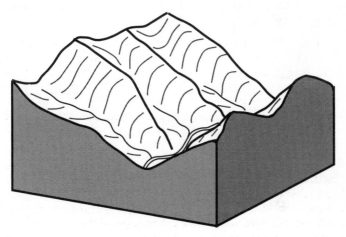

figure 8.3 A landscape sculpted by the action of rivers and streams. If the main valley were deepened and widened by a glacier, the landscape could end up like that shown in Figure 8.2.

Drainage patterns

So far we have been talking in generalities. In detail, how a landscape is eroded depends as much on the nature of the bedrock as on the agent of erosion. This is particularly seen in the erosive effects of water. If the rock is limestone, which is soluble, rainwater may drain through a system of fissures and caves, and there may be no streams at the surface at all for most of the year. If the rock was laid down in a sequence of sedimentary beds, the softer units will be worn away most rapidly. If there is a strong pattern of faulting or jointing running through the rock, rivers and streams will tend to exploit and widen these lines of weakness. The nature of the bedrock, and its geological structure, can therefore often be inferred from a map showing the drainage pattern (Figures 8.4 and 8.5). Catastrophic floods released by retreating glaciers create a valley pattern at odds with 'steady-state' conditions; the channelled scablands referred to above are crossed by flat-bottomed, near vertically sided canyons up to 200 metres deep, carved into the Columbia River flood basalts.

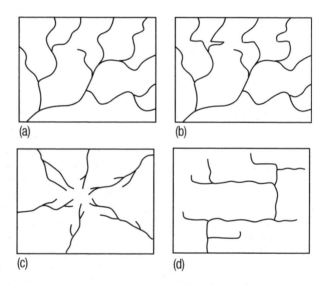

(a)   (b)

(c)   (d)

figure 8.4 Varieties of drainage pattern: (a) dendritic, which develops where the bedrock is fairly uniform; (b) a dendritic pattern offset by recent fault motion – in this case movement on an east–west fault has offset the drainage in the northern region to the left. If the fault remains stationary for long enough, the drainage pattern will readjust; (c) radial drainage, developed over a central high point, such as a domal uplift, or a volcano; (d) rectangular drainage, which exploits planes of weakness such as joints or softer beds of rock.

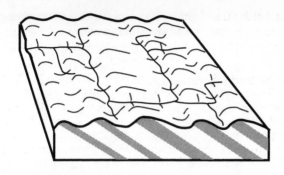

figure 8.5 Block diagram showing how the rectangular drainage pattern in Figure 8.4(d) can be related to the underlying geology. The structure is simple, with alternating harder (shaded) and softer beds dipping to the right. Streams develop along valleys where the softer rock has been worn away at the surface, and occasionally cut gorges at right angles to these, through the intervening ridges of harder rock.

Marine erosion

Finally we should look at erosion by the sea. On coastlines exposed to storms, waves can be a very powerful erosive agent, and can cut the coastline back into a cliff. Erosion is strongest near the level where waves are breaking, so it does not cut down much below the intertidal level. A wave-cut platform of planed-off rock is often exposed at low tide, extending seawards for 100 metres or more from the foot of the cliff, which can make for an enthralling setting to explore the well-exposed rock sequences and to search for any fossils within. There is no comparable limit to the height above sea level of the top of a cliff, this being controlled solely by the height of the land surface.

Where the rock is a sediment composed of gently dipping layers, a cliff is liable to be undermined, particularly if the beds at the foot of the cliff are relatively soft rock. In such circumstances it is common to find caves at the cliff foot. Faults or joints can be exploited too, and these often control the locations of coves and bays. As the coast retreats, a headland that once separated two bays may become a steep-sided island or sea-stack. Some new islands of this sort inherit a cave running through them and are known as sea arches. Figure 8.6 illustrates successive stages of coastal erosion. Coastal retreat on this scale takes thousands to hundreds of thousands of years, depending on the strength of the rock and the violence of the storm waves.

In Figure 8.6, the coastline is likely to be marked by cliffs, except within the bays, which are sheltered from the strongest wave action and where sand is likely to accumulate. You will consider the circumstances under which sediment can be deposited, and the sort of rocks which these can eventually form, in the next chapter.

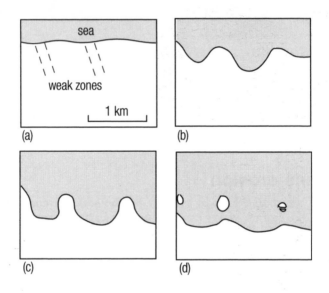

figure 8.6 Maps of successive stages (a–d) of coastal erosion. In (a) the coastline has weak zones running into it. These may be faults, closely spaced joints or beds of softer than usual rock. In (b) and (c) marine erosion has eaten away at the weak zones fastest, resulting in a coastline of headlands and bays. In (d) the headlands have become small islands.

09

deposition of sedimentary rocks

In this chapter you will learn:

- how beds of sediment often contain information about the currents that deposited the grains
- about the types of sediment that accumulate in different environments
- how sediment can be turned into rock.

The most hospitable landscapes, including the most profitable farmland in most countries, are usually in regions where rock is being deposited in the form of sediments, rather than being eroded. Naturally, these tend to be low-lying areas of land. Much more sediment is deposited, unseen, below the sea.

Sediments tend to be deposited in layers, known as **beds**, so, after they have hardened, they form the most easily worked building stones. They are sources of certain vital resources, such as coal and oil, and may act as reservoirs for underground water supplies. Sediments can accumulate to great thicknesses, for example the Grand Canyon of the Colorado River in Arizona has cut through 2 km of sedimentary rocks, which were deposited in several episodes between about 550 and 250 million years ago.

We have already considered the processes involved in erosion and transport of detrital grains. In this chapter we will begin by looking at the conditions necessary for grains to be deposited. Many of the most familiar sedimentary rocks are formed from accumulations of detrital material. These are known as **clastic** rocks, **clast** being a term used to describe a fragment or grain. Clastic sedimentary rocks are classified by the diameter of their most abundant clasts according to the scheme in Table Y of Appendix 2 (page 257). You should refer to this if you are concerned to know the exact size distinctions that are drawn between clay particles, silt, sand, gravel and so on.

Most deposits of sediment contain only a limited range of clast sizes. The reason for this is that a current of wind or water that is just weak enough to deposit clasts of a particular size will not be flowing strongly enough to bring in clasts that are much coarser than this. However, it will be flowing with sufficient speed and turbulence to keep all the significantly finer clasts in suspension, so those will not be deposited in the same place.

Ripples and dunes

Except in very quiet conditions, when fine silt and mud can settle out from the suspension-load, most clastic sediments are formed by deposition from the bed-load. The rolling and bouncing (saltation) modes of transport in the bed-load give rise to some distinctive features in the bed of the flow, which can very often be recognized in ancient deposits. Everyone is aware that wind tends to pile sand into large dunes. Water flow produces similar but usually smaller sedimentary structures, of which ripples are the best known. Ripples, dunes and the like are referred to collectively as **bedforms**.

Many ripples and dunes are characterized by asymmetric cross-sections. Each has a gentle slope on its upstream side and a steeper slope on its downstream, or lee, side. Grains bounce or roll up the gentle upstream slope, which is exposed to the full force of the current. As they topple over the crest they find themselves in a more sheltered region, where the current is too weak to transport them, so they simply drop onto the upper part of the lee side. This face is continually being over-steepened by the arrival of sediment at its top. Every time it gets too steep to be stable, an avalanche of grains slips down the face. Each avalanche forms a layer resting on this slip-face at the maximum angle that can be supported, which is around 30° or 40°. Grains are thus removed from the up-current side of the bedform and deposited on its lee side, and so the ripple or dune migrates slowly downstream.

Figure 9.1 shows how bedforms like ripples and dunes migrate.

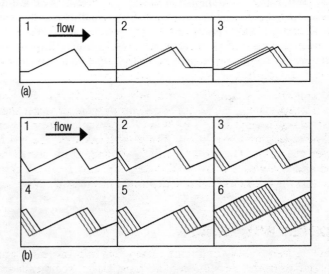

(a)

(b)

figure 9.1 Migration of ripples (centimetres in size) or dunes (tens or hundreds of metres in size) seen in cross-section. In (a) a single ripple or dune is shown, in a situation where there is no net accumulation of sediment. The three views are at successive times, with the instantaneous surface shown in the black solid line and previous surfaces in the faint grey line. In (b) there is net accumulation, so that each bedform climbs up the back of the previous one as it migrates. 1–5 are successive time steps, and 6 is much later.

If there is no net deposition of sediment, all that happens is that bedforms migrate downstream. However, if more sediment arrives in the area than can be transported away, then there is net deposition. Successive bedforms migrate up the upstream faces of previous ones, and a deposit of sediment accumulates. The successive slip-faces within each bedform can often be recognized if the deposit is subsequently excavated and seen in cross-section, even after hundreds of millions of years (Figure 9.2). Slip-faces dip more steeply than the deposit as a whole, and when this is seen the contrast in dip amounts gives rise to a feature commonly called **cross-bedding** (Plate 7).

When conditions are such that more sediment is being removed from an area than is being deposited, migrating bedforms may still develop, as can be seen in modern environments. However, for obvious reasons, evidence of this tends not to be preserved in the geological record.

The shapes of bedforms as seen from above depend on the conditions under which they form. A relatively slow flow in a constant direction produces transverse ripples or transverse dunes, with their crests aligned at 90° to the current. As the flow becomes stronger, crests become wavier, eventually breaking into crescent shapes. These contrasting bedforms are associated with different patterns of cross-bedding that allow them to be distinguished when seen in cross-section, as shown in Figure 9.3.

figure 9.2 The face of this sand quarry in County Durham, England, shows cross-bedding formed in desert sand dunes about 250 million years ago.

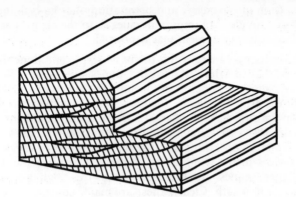

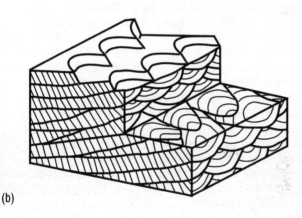

figure 9.3 Two sorts of cross-bedding, both produced under conditions of net accumulation of sediment. Each diagram is a block view, cut on flat horizontal and vertical faces to show how the cross-bedding appears on different planes. The actual surface, showing the three-dimensional shape of the bedforms, is the left-hand part of the top of each view: (a) transverse ripples or dunes, (b) crescent-shaped ripples or dunes. The cross-bedding is different in each case: (a) has tabular cross-bedding, (b) has trough cross-bedding. It would be difficult to distinguish between these if the only cross-section you saw was looking at right angles to the current direction (the left-hand face), but the distinction is clear when looking along the current direction (the right-hand face).

Observation of the nature and orientation of cross-bedding can therefore provide very important clues as to the environment in which an ancient sediment was deposited.

There are many variations on the basic ripple and dune theme. Crescent-shaped wind-blown sand dunes are usually isolated from one another, rather than touching at their tips as suggested in Figure 9.3(b). These are called barchans. In some areas dunes develop with their crests aligned parallel to the dominant wind direction; these are known as seifs or longitudinal dunes. They have slip-faces on both sides, usually attributed to corkscrew-like vortices in the wind as it is funnelled between the dunes. The largest dunes are hundreds of metres high and many kilometres in length. Clearly these are major constructs and are not going to re-orient themselves every time the wind changes direction!

Most dunes migrate by about a metre per year. The shape of a dune, and the internal cross-bedding it inherits, are influenced not just by the annual prevailing wind but also by seasonal winds blowing from other directions. The surfaces of sand dunes are often covered in transverse ripples, and these are small enough to adjust in response to hourly variations in wind strength or direction.

Ripple-like bedforms often develop within pyroclastic flows. Here, despite the high speed of the flow, the high ratio of the pyroclastic fragments to the volume of air entrained within the flow is so great that clasts are plastered onto the ground even when the speed of the flow would otherwise be great enough to keep them moving in the bed-load. Pyroclastic flow deposits, and also airfall ash, can be considered as sediments even though they have a volcanic origin.

Turbidity currents

There are subaqueous analogues to dense pyroclastic flows, known as turbidity currents. These are particularly important for transporting sediments off the edge of the continental shelf and onto the ocean floor, notably into trenches. What happens is that a submarine landslide stirs up a lot of sediment into suspension. This mixture is considerably denser than the surrounding seawater, and flows downslope as a turbulent mass of sediment and entrained water.

Turbidity currents can be very powerful, and have been known to snap submarine telephone cables. Initially, a turbidity current is usually strong enough to scour up extra material at its base, but as the slope eases, it slows down and begins to deposit its load. At first, deposition is by plastering down of clasts as in pyroclastic flows, but as the speed wanes and the proportion of sediment remaining in the turbidity current decreases, ordinary climbing ripples form. These can be distinguished from ripples formed in more familiar water currents only by the context in which they are found. After a turbidity current has passed by, the fine-grained material that had been stirred up into suspension settles out, so the sandy rippled part of the deposit is overlain by finely laminated silts and muds. A deposit laid down by a turbidity current is called a turbidite.

Current directions

In rivers, the direction of the current is rarely as fickle as the winds, and is essentially uniform. Even though the strength of the current may vary seasonally, the consistency of current direction controls the orientation of the cross-bedding so that unidirectional current flow can be recognized in ancient river deposits.

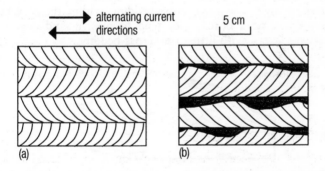

figure 9.4 Cross-sections showing ripple cross-bedding produced by alternating tidal currents. In (a) the top of each layer of ripples has been planed off (eroded) by the current that brought in the sediment that formed the next layer of ripples. In (b) there is a quiet interval between each reversal of the tide, allowing mud or silt (black) to accumulate in the hollows between ripples. This is called 'flaser bedding'.

In the sea, tidal currents change in both strength and direction. In sediments deposited in shallow marine conditions, where tidal currents are most obvious, this is usually reflected by alternating directions of cross-bedding, and sometimes by thin layers or lenses of fine-grained suspension-load that was able to settle during the period of slack water while the tide was turning (Figure 9.4).

Waves

At sea or in large lakes where water depth is less than about ten times the height of the waves, wave motion can disturb the sediment. Bedforms may grow in the shape of symmetric ripples whose opposite faces have equal steepness, quite unlike the asymmetric ripples produced by current flow (Figure 9.5). If a current is flowing in a totally different direction from the alignment of wind-generated waves, differently oriented current ripples and wave ripples may produce a criss-cross pattern of interference ripples. These can often be seen exposed on modern beaches at low tide. During storms, wave action may be effective at many tens of metres depth. Particularly violent storms are thought to be the cause of what is known as hummocky cross-stratification, when hummocks form that are a few metres across but only a few centimetres high, with correspondingly low-angle cross-bedding within them. When waves break on a beach, they tend to plaster sand down in successive layers, with little or no cross-bedding.

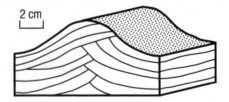

figure 9.5 Cross-sections through a ripple produced in shallow water as a result of wave motion.

Finer and coarser sediments

Ripples and dunes are usually found only in sand or the coarsest silty sediments. Fine silt or mud will not settle out from water unless it is flowing at less than a few centimetres per second, which is too slow to sculpt sediment into bedforms. Silt and mud thus tend to accumulate in fine horizontal layers, referred to as laminations.

Conversely, when gravel and pebbles are deposited by flowing water, it is usually because the current has dropped suddenly, which leaves too little time for bedforms to develop. However, pebbles do tend to be oriented by the current and come to rest with their tips pointing downward into the current. Each pebble thus overlaps its neighbour, in a fashion described as imbrication (Figure 9.6). This can be a useful indicator of current direction. It is virtually unknown to find an ancient deposit containing nothing but pebble-sized clasts, because finer material (sand or mud) is likely to be washed in later, filtering its way down through the gaps.

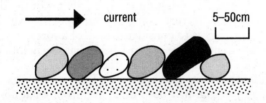

figure 9.6 Cross-section showing imbricated pebbles deposited by fast-flowing water. The tip of each pebble points upstream and slightly downward.

Other means of transport

We have seen previously that material may also be transported by mudflows (lahars), by landslides and by moving ice. After this material has been deposited, it is generally distinguishable from a water-laid deposit because of its wider range of clast sizes, the angular, poorly rounded shapes of the clasts, and the lack of bedforms. Another distinguishing feature is that pebble-sized clasts may be supported by a matrix of mud or sand, whereas in a water-laid deposit pebbles almost always touch.

Deposition from mudflows and landslides occurs simply when the flow stops moving. Deposition from a glacier happens as the ice melts, usually at its downhill end, or 'snout'. Often, glacial detritus is reworked by streams of meltwater, in which case it loses all or most of the characteristics listed above. Debris surviving more or less where it was dropped by a glacier, without being reworked, is referred to as moraine. When the melting snout of a glacier has retreated up a valley in stages, each successive position of the snout is marked by an accurate ridge of moraine.

Using sediments to interpret ancient environments

Most sedimentary rock was deposited under water. We have already examined the separate processes that are involved, and the bedforms that can be created. However, to interpret the environment in which an ancient deposit was laid down it is rarely sufficient to rely on such a tightly focused approach. Usually, it is better to weigh up many attributes of the deposit, and to build up a picture using several lines of evidence. A rather good analogy is to consider the job of a forensic scientist, called to the scene where a dead body has been found. The police would be unimpressed if the forensic expert merely reported that the victim had been shot. They would expect to be told what sort of gun had been used, and from what angle and direction. Moreover, they would want to know how long the body had lain there before discovery, whether the body was found in the place of death or had been moved after death, whether or not there were signs of a struggle, and so on. The job of a geologist interpreting a sediment is much the same.

To see how the evidence preserved in the sedimentary record can be pieced together to interpret past environments, we will look at a few key modern depositional environments. By establishing the ranges of bedform, grain size, composition and texture, and their lateral and vertical variations and extent, we can understand what to look for in ancient deposits. This is a principle well known among geologists, which can be summed up as 'the present is the key to the past', an expression attributed to Sir Charles Lyell (1797–1875), one of the fathers of modern geology. It does not always work – for example there is nothing like channelled scablands forming today, and it took the geological community 50 years to accept that they had been formed by a catastrophic process after the idea was proposed in the 1920s – but it remains a trusted approach.

Depositional environments associated with rivers

We will look first at rivers, in their sluggish downstream reaches where they are depositing rather than eroding material. Unless it has been straightened artificially, a lowland river usually winds in a lazy way across its **flood plain**. This flood plain consists mostly of muds and silts deposited as a result of floods, when the rate of water supply from upstream was so great that the flow could not be contained within the channel. While the flood is abating, the water covering the floodplain is sufficiently stagnant for most of the suspension-load of silt and clay particles to settle out, although normally (when the river is confined to its channel) this would be flushed right down the river until it reached the sea or a lake. A flood plain is thus built of successive layers of fine- or very fine-grained flood deposits, which provide some of the globe's richest agricultural land. On the other hand it is a foolish place to build houses because after exceptional rainfall it is vulnerable to flooding, unless there are expensive and well-maintained 'flood defences'.

Slightly more sediment is deposited on the flood plain immediately next to the river, and in this way riverbanks are built up, in some cases to heights of several metres. These raised banks, known as levées, are a characteristic feature of rivers such as the Mississippi.

The river itself usually flows too fast for its suspension-load to be deposited, and is likely to be transporting sand in the bed-load. We therefore find ripples when we look on the bed of a river; however, these can be preserved in the geological record only if they are buried by later deposits. Burial of ripples can happen without clogging up the river, because a winding lowland river does not maintain its course. The current has to flow faster round the outside of a bend than on the inside. The riverbank on the outside of a bend is therefore subject to undercutting and erosion. In contrast, bed-load tends to be deposited against the bank on the inside of a bend, where the current is slowest, producing a sand and gravel deposit known as a point-bar. The effect of this is for bends, or meanders as they are known, to become progressively exaggerated, as shown in Figure 9.7. A meander loop may eventually become cut off, allowing the main course of the river to straighten again, but another meander soon develops and the net effect over tens of thousands of years is for the river to migrate from side to side across its flood plain.

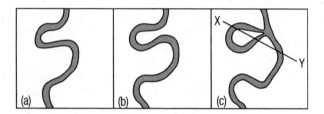

figure 9.7 Map view at successive times showing how a meandering river migrates across its flood plain, by erosion on the outside of bends and deposition on the inside. In (c) a loop has been virtually cut off, forming an oxbow lake that will eventually silt up. X–Y indicates the cross-section shown in Figure 9.8.

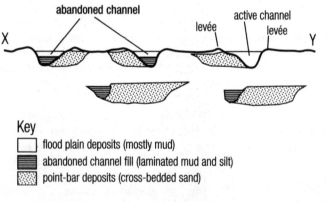

figure 9.8 Cross-section along the line X–Y in Figure 9.7, showing lateral and vertical variations in the nature of the deposits. The channels at ground level are those seen in the map view. The buried channels mark the course of the river at earlier times. These were abandoned and then buried by flood-plain deposits.

The kind of thing a geologist would hope to see in a cliff cut into an ancient flood-plain deposit can be pictured by constructing a cross-section across a modern environment, such as Figure 9.8. Most of the deposit is muds and fine silts, laid down on the surface of the flood plain. Abandoned channels become filled with laminated silts and muds, looking rather similar to flood-plain deposits except they will have steep edges marking the sides of the channel where it was cut into the flood plain. Successive positions of active channels are marked by point-bar accumulations of sand and gravel deposited on the inside of bends. The extent of each such sand/gravel sheet depends on how far the bend migrated before the channel was abandoned. Within each point-bar deposit, the grain size becomes finer upwards, with coarse sand or gravel at deeper levels, representing a faster current, and finer sand towards the top where the current was slowest. Ripple cross-bedding indicates the direction of the current.

Sediments laid down by a meandering river thus consist predominantly of very fine-grained material (mud and silt), most of which was laid down on the flood plain. However, there are sheet-like or lens-like bodies of gravel and sand within it that provide the evidence for migrating channels.

Not all rivers have this sort of flood plain. Downstream from the ends of glaciers, or in regions where rainfall is strongly seasonal, rivers have to carry a particularly high sediment load at certain times of the year. Their downstream reaches are not confined within a single channel. Instead, the river occupies

figure 9.9 Map view of a braided river, which may be 100 metres to about 10 km in width. Braided rivers are common in Texas (the Brazos is a classic example) and other areas where discharge is strongly seasonal.

several interconnecting channels, separated by bars of gravel that are submerged only when the river is in flood. This is known as a braided river (Figure 9.9). Seen in cross-section, the deposits from a braided river consist almost entirely of lenses of sand and gravel (Plate 8). Any mud is restricted to the occasional abandoned channel, so it is relatively easy to distinguish between deposits from meandering and braided rivers.

Flash floods

The mountainous fringes of desert areas experience even more strongly episodic river flow. Here, channels may be dry for almost the entire year, or even for decades. However, when it does rain it usually does so very heavily, and gullies that are almost always dry may rapidly become brim-full with water flowing fast enough to roll even quite large boulders along in the bed-load. A flash flood such as this is one of the chief hazards to be aware of when working in mountainous terrain. Where a gulley debouches onto the plains, the torrent is no longer confined and so it disperses. This causes the speed of flow to drop dramatically, so the larger clasts in the bed-load come to rest. The deposit spreads out radially from the mouth of the gully, to form an alluvial fan. Seen in cross-section, it consists of successive layers of imbricated boulders and pebbles, with sand washed in between later (Plate 9). Channels within it may not be obvious but, with care, erosion surfaces can usually be distinguished at the base of each fresh influx of sediment.

Deltas and estuaries

Turning our attention now to where rivers reach the sea, we often find a delta made of sediment deposited by the river. There are two factors contributing to this. The first is that, just as in an alluvial fan, where the flow spreads out its speed drops, so any bed-load that the river has managed to bring all the way to the sea will now be deposited. The second is a result of something that happens to clay particles as soon as the water becomes salty. Clay particles are tiny, and remain in suspension provided they stay as individual particles. However, once they get into even slightly salty water they start to stick together, or 'flocculate'. The clusters of clay particles created in this way are big enough to sink, and so the suspended load of the river is dumped as soon as the river water mixes with the sea water.

The world's two most famous deltas show some important differences in morphology, resulting from differing local conditions (Figure 9.10). The archetypal delta is that of the Nile. The likeness between the shape of this flat, many channelled tract of land to the Greek capital letter delta, Δ, led the geographer Herodotus to coin the term 'delta' to describe it in 490 BC. The Nile breaks into a plethora of distributory channels, radiating outward from the vicinity of Cairo, more than 100 kilometres from the sea. The seaward fringe of the delta is a gentle arc, rather than the straight edge implied by the Greek character. In contrast, the Mississippi delta has been likened to a bird's foot in appearance, because the levée-confined channel that feeds it has built out beyond the main shoreline and it and the few distributory channels coming from it stick out into the sea like the toes on a foot.

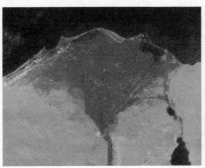

figure 9.10 Views from space of two famous deltas. On the left is a 200-km-wide view of the classic delta of the river Nile, which flows from south to north. Sea is dark, desert is pale, and the well-vegetated land on the delta is mid-grey. On the right is a 80-km-wide view of the 'bird's foot' delta of the Mississippi, which flows from north-west to south-east. Deep water is dark, shallow water is light, and the complexly shaped land is in mid-tones.

The main reasons for the different shapes of these two deltas are that the Mississippi discharges much more sediment than the Nile, whereas the Nile delta is exposed to stronger waves, which tend to mould a delta front into an arc.

Neither the Nile delta nor the Mississippi delta is subject to strong tides. Where there is a large tidal range, for example at the mouth of the Ganges-Brahmaputra or the Mekong, tidal currents scouring up and down the distributary channels flush a lot of the sediment seawards, and keep the channels wide. Other river mouths lack any kind of delta, but are funnel-shaped inlets of the sea known as estuaries. Most of the river mouths around the British Isles and on the east coast of North America from Chesapeake Bay northward are of this kind. Factors encouraging the presence of an estuary rather than a delta include relatively low amounts of sediment discharge and high tidal ranges. Some estuaries are valleys that have been partly submerged by the rise in sea level that happened as a result of the melting of the main ice sheets at the close of the most recent glaciation.

In many ways the tops of deltas resemble flood plains of meandering rivers. The deposits found there consist of ripple cross-bedded sand that accumulated on point-bars, silts and muds laid down in abandoned channels, and mud deposited over the delta plain during floods. Delta plains are often densely colonized by vegetation, and roots, fallen leaves and branches can often be found as fossils. Decaying vegetation may accumulate thickly enough to form coal deposits, as you will see in Chapter 11.

Moving seawards from a modern delta, the sandy sediment deposited near the shoreline may be reworked by wave action, which is reflected in the style of cross-bedding. This area is referred to as the delta bar; channels are largely absent here, and so is mud because such shallow water is usually too agitated for clay to settle, even when flocculated. Another diagnostic feature is broken or complete shells of marine organisms, which are, of course, absent in any sediment deposited higher up the river. Going slightly further offshore, to the delta front, we reach the zone where most of the flocculated clay is able to settle. This region is characterized by unbroken shells of marine organisms, and burrows in which these and soft-bodied organisms lived.

Prograding deltas

We can now introduce a very important concept. By their very nature, because sediment is continually being brought there, deltas tend to build out seawards over time. A shoreline advancing seawards like this is said to be **prograding**. As a result, if we follow events at a specific point, originally a few kilometres offshore, over maybe 10 000 years we might record the following sequence. Purely marine deposits are encroached upon by delta-front sediments. At first these will be the very finest clay material, which has been washed furthest from the delta, but as the delta draws nearer coarser clay and silt are deposited here. Throughout this time the seabed is getting higher, because of the accumulation of sediment. Eventually we will find ourselves on the shoreline, where the sands of the delta bar are being deposited. The shoreline continues to prograde seawards, and we find ourselves in a distributary channel of the delta, whose base may erode away some of the underlying delta bar sands. After a while our channel migrates away, and we find ourselves on the delta plain, though there is always a chance that a channel may migrate our way again if we wait long enough.

Those without the patience to wait 10 000 years could see the same variety of environments by moving sideways a distance of ten or so kilometres from offshore to delta top. We could also see the same sequence of environmental changes by drilling a 100-metre-long vertical borehole through the top of a modern delta. This would reveal delta plain sediments at the top, delta bar sediments below, and progressively finer delta-front

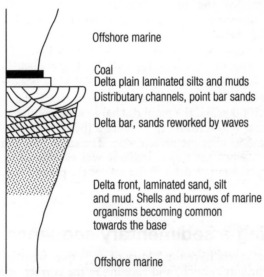

Offshore marine

Coal
Delta plain laminated silts and muds
Distributary channels, point bar sands

Delta bar, sands reworked by waves

Delta front, laminated sand, silt
and mud. Shells and burrows of marine
organisms becoming common
towards the base

Offshore marine

figure 9.11 Changing grain size and bedforms in a vertical sequence through a prograding delta. The width of the column is drawn to correspond with changes in grain size. The basic sequence, fine grained (marine) at the bottom, becoming gradually coarser upwards until topped by non-marine muds, is characteristic of deltas. The relative contributions of waves, tides, river flow and sediment supply influence details in this general picture.

sediments below these (Figure 9.11). In other words, the lateral juxtaposition of sedimentary environments at an instant in time is reflected by environmental changes seen in a vertical sequence (representing the same location at over a period of time). This relationship is known to geologists as Walther's Law, after Johannes Walther who first stated it in 1894.

Figure 9.11 shows the delta plain sediments overlain by a new set of offshore marine sediments. This might surprise you, because continuing our imaginary sojourn on a delta, you might expect that eventually we would find ourselves well and truly on dry land as the shoreline continues to prograde. This may be the long-term tendency, but it is quite likely that we will see several returns to marine conditions in the meantime. This is because if our river bursts through one of its levées near the upstream part of the delta, the main discharge may be diverted into a new channel, in which case a new area of delta will build out into a different area of the sea. This will starve our observation point of sediment, which will slowly sink below sea level as a result of compaction of the underlying sediments, and downward isostatic adjustment of the crust as a whole because of the weight of sediment.

To take the example of the intensively studied Mississippi delta, we know that the present-day main channel (seen in Figure 9.10), flowing south-eastwards to the sea from New Orleans, came into existence only about 1000 years ago. For most of the previous 5000 years, the main discharge went southwards from an area near Baton Rouge to produce a (by now mostly sunken) delta to the west of the present one. The same period also saw an interval when the main discharge was eastwards from New Orleans, producing a delta to the east of the present one, part of which is still above sea level.

Logging a sedimentary sequence

Geologists rarely have the luxury of cliffs that are high enough and continuous enough, and running in the correct orientation to record the whole lateral extent of a delta. However, by interpreting the environmental changes seen in a single quarry face, or in a core recovered from a borehole, or in a series of isolated exposures, they can construct a log such as that in Figure 9.11 and then use Walther's Law to deduce what the whole area would have looked like at the time when the sediments were being deposited.

Beaches and other coastal environments

The concept is not just applicable to deltas; it works whenever there have been gradual changes in sedimentary environment. For example, a sedimentary sequence deposited over time at a place where the shoreline was prograding because of accumulation of sand washed onto a beach would begin at its base with marine deposits characteristic of deepish water, below the level exposed at low tide, and too deep for storm waves to disturb. These sediments, deposited in conditions of unvarying low energy, would be very fine grained, probably either mud or limestone (which we will discuss shortly). The slow accumulation of this fine-grained deposit would eventually bring the sea floor close enough to the surface of the water for storm waves to disturb it, and we might expect to see erosion surfaces and sandy layers, possibly containing hummocky cross-stratification representing big storms.

As the sea floor became higher and the general environment conditions became more energetic, with stronger tidal currents and disturbance by waves at every low tide, mud or limestone would be replaced by silts and fine sands. Such conditions are favoured by many burrowing organisms that feed off the continual supply of nutrients brought in by the currents. Cross-bedding, such as wave-generated ripples, may be disturbed or obliterated by burrows. We might also find complete fossilized shells or organisms preserved within their burrows.

Continuing up the sequence we would reach sand deposited on the beach, between the low and high tide marks. This is generally laid down in low-angle sheets with occasional ripple cross-bedding. There would probably still be plenty of shelly material, but now it would be mostly broken fragments. Sand on a modern beach is usually unlike river sand because although quartz may be the dominant mineral, a large proportion of the sand grains are actually fragments of broken shells and so are made of calcium carbonate. If our beach were the sort with sand dunes behind it, the top of our sequence would contain the large-scale cross-bedding that is characteristic of wind-blown dunes. Because the sand had not been blown very far, the grains would be unlikely to have become very well rounded and frosted, so would not resemble the grains typical of desert sand dunes. Eventually, beach dunes become colonized by plants, and so we might find a few rootlets in our deposit.

The succession of sediments we have described would be most likely to be preserved in the geological record if it sank below sea level and was buried below another cycle of prograding shoreline sediments. Otherwise, it might be eroded away by river action, or by a change in marine conditions, For example, if rather than bringing a net supply of sand to the beach, the currents started to take more sand away than they brought in, the beach and eventually the dunes behind it would be washed away.

There are many variations on the theme of coastal sediment deposition. If the tidal range is great and the gradient gentle, waves break a long way offshore, and so lose most of their energy by the time they lap up to the high tide mark. Particularly on sheltered coastlines, there may be a sand beach exposed at mid-tide, backed on the landward side by an extensive mudflat consisting of very fine-grained material that settles out in quiet, wave-free conditions while the tide is turning. Today in temperate regions, these tidal flats become colonized by coarse grasses to form salt marshes. The grass roots bind the sediment together, preventing it from washing away and encouraging it to build up above the normal high tide level. In humid tropical regions, the same function is served by mangrove trees, whose roots spread out above ground level and act as a trap for sediment.

Evaporite deposits

In arid regions there is insufficient rainfall to support salt marsh or mangrove swamp vegetation. Instead, above the normal high tide mark, mud is matted down by layers of salt-tolerant fibrous algae. Such a salt flat is known by its Arabic name of 'sabkha'. The mud here is not dominated by clay minerals but by mud-sized grains of carbonates, chiefly calcite, $CaCO_3$, and dolomite, $CaMg(CO_3)_2$.

There are biological ways of growing these carbonates that we will look at shortly. However, on a sabkha they form inorganically. The sun beating down on the surface of the sabkha causes evaporation, and seawater is continually drawn in through the mud to replace what is lost. As the water evaporates, its dissolved constituents come out of solution, often beginning with carbonate minerals. Minerals formed in this way are described as **evaporites**. If the rate of evaporation is sufficient, other evaporite minerals begin to grow, notably gypsum ($CaSO_4.2H_2O$) anhydrite ($CaSO_4$) halite (NaCl, which is table salt), and ultimately salts of potassium. Gypsum deposits make alabaster, much used by sculptors and from which plaster of Paris is made. Table salt, of course, has many uses.

Evaporite minerals do not only form in sabkhas. They can also form at sea, provided evaporation of seawater is proceeding faster than the rate of replenishment by fresh water. Such conditions are met in enclosed marine basins in sub-tropical latitudes where there is little rainfall. Here, the seabed becomes covered by crystals of evaporite minerals that have grown near the surface, where the effects of evaporation are most extreme, and sunk to the bottom. There are no extensive marine evaporite deposits forming today, but there are up to two kilometres of evaporites on the floor of the Mediterranean, deposited between 5.4 and 4.8 million years ago when the Mediterranean was periodically cut off from the Atlantic by the emergence of the Straits of Gibraltar above sea level (Figure 9.12). To produce such a thickness of evaporites requires the Mediterranean to have been completely evaporated about 40 times. Even today, the rate of surface evaporation from the surface of the Mediterranean Sea exceeds the rainfall and river influx. If the Straits of Gibraltar became closed, preventing replenishment by Atlantic water, the Mediterranean would take only about 1000 years to dry out.

figure 9.12 These large 'swallow tail' crystals of the evaporate mineral gypsum, today exposed on the uplifted island of Cyprus, grew on the floor of the dessicating Mediterranean Sea about 5 million years ago.

Economically important evaporite minerals that formed in an enclosed marine basin known as the Zechstein Sea about 220 million years ago are found in a deposit stretching from north-east England to northern Germany. These include potassium salts, and are the foundation of the British chemical industries of Teesside.

Limestone

Evaporites form only under special circumstances. However, calcium carbonate is the main rock-forming mineral in many sedimentary environments. It owes its ubiquity to biological, rather than physical or chemical processes. A rock made mostly of calcium carbonate is called limestone, which is by far the most abundant non-silicate rock type. Deposition of calcium carbonate is an important part of the rock cycle, in which calcium that was dissolved during weathering and the bicarbonate by-product of hydrolysis come out of solution in tandem, maybe thousands of kilometres from where the calcium was dissolved. Calcium carbonate is most commonly found as the mineral calcite.

Sea shells found on the beach are made of calcium carbonate, and some limestones are made of deposits of such shells, or of shell fragments. More important globally are microscopic, generally single-celled plants and animals that live in the sea. Most drift around near the sea surface where there is plenty of sunlight. These are referred to as plankton. Less abundant forms live on the sea floor. The most important animal forms are foraminiferans, which are simple organisms similar to the amoeba except for having multi-chambered calcareous shells, commonly up to about a millimetre across. The most important plant variety is the coccolithophores. These are smaller and rather than having a simple shell they are covered in an array of disc-like plates called coccoliths, typically only 20 micrometres in diameter, that separate after death.

When microscopic plankton like these die, their shelly remains sink towards the sea floor. If conditions are gentle enough, they can settle to form extensive deposits. The famous White Cliffs of Dover are cut into a 70-million-year-old accumulation of coccoliths that was deposited on the floor of a calm, shallow sea that had no major rivers discharging muddy sediments into it (otherwise the cliffs would not be so white). This rather special kind of limestone is called chalk.

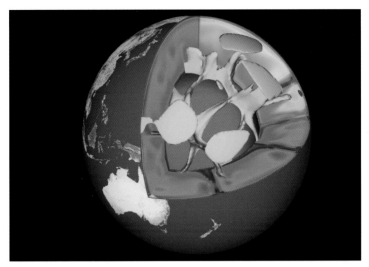

plate 1
Possible mantle convection pattern below the Pacific. In the cross-section views, yellow and orange represent areas of slightly warmer rising mantle. Slightly colder sinking zones are shown in purple. In the cut-away portion, only the rising plumes are shown. The rest of the mantle has been made transparent, to reveal the core within.

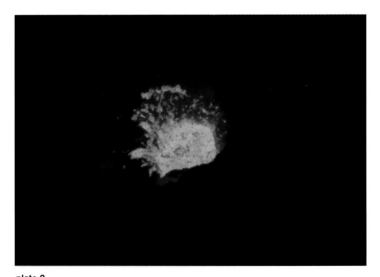

plate 2
Magma that has reached the surface at a several metre-wide vent inside the active crater of the volcano Masaya in Nicaragua, seen at night. A few seconds before this picture was recorded, a small explosion distributed blobs of molten lava beyond the rim of the vent. These are beginning to cool down, and are glowing less strongly than the 1100 °C magma inside the vent.

plate 3
Explosive eruption on Mount Etna volcano. A sustained fountaining event is sending incandescent material to a height of about a hundred metres above the rim of the crater, whereas a separate explosion has produced a smaller cloud of grey ash directed obliquely towards the viewpoint.

plate 4
Vertical sheets of basalt (dykes) feeding overlying pillow lavas, in the Troodos ophiolite, Cyprus. This is crust created by sea-floor spreading on an ocean floor. The major dyke on the right is about a metre wide.

plate 5
Slowly spreading pahoehoe lava on the flanks of Kilauea volcano, Hawaii.

plate 6
Close-up view of a granite. Pink and white crystals are two varieties of feldspar mineral, clear, grey, crystals are quartz, and black flecks are biotite. The brassy metallic minerals are sulfides of copper (pyrites and chalcopyrites), and the metallic mineral with a bluish sheen is molybdenum sulphide (molybdenite).

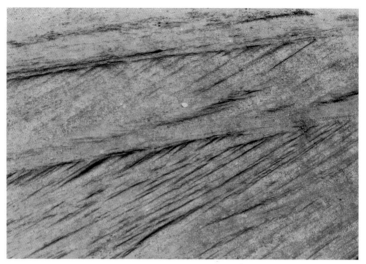

plate 7
Cross-bedding in sandstones deposited in a braided river, approximately 300 million years ago. The coin provides a scale. (Roaches Grit, Staffordshire, England).

plate 8
The upper two-thirds of this sandstone cliff shows cross-sections through a series of river channels, each containing the characteristic cross-bedding of point-bar deposits. (Kalbarri, Western Australia).

plate 9
The deposit of small boulders was dumped here when the force of a flash-flood was spent. (Wadi Ahin, Sultanate of Oman)

plate 10
Complexly deformed gneiss, in which the original metamorphic layering has been folded and then re-folded to produce the pattern visible on this surface. The pale quartz-filled vein to the left of the hammer was evidently injected after the last phase of folding. (Yilgarn block, Western Australia)

plate 11
Reconstruction of the swampy conditions that led to the formation of coal in Carboniferous times. The vegetation is of types known mainly from fossils. Dragonfly in foreground has a wingspan of about 60 cm.

plate 12
An open-cast copper mine, whose ore is a result of hot fluids circulating through the ocean crust near a spreading axis. This pit is more than a kilometre across. (Skouriotissa, Cyprus)

plate 13
A 10 cm long piece of limestone showing abundant internal and external fossil moulds of bits of crinoid stem. (Cromford, Derbyshire, England).

plate 14

This 1-metre-high miniature cliff could be on Earth, but in fact it is on Mars, as seen by a NASA rover. The cross-bedding visible here was probably formed by sand transport in water.

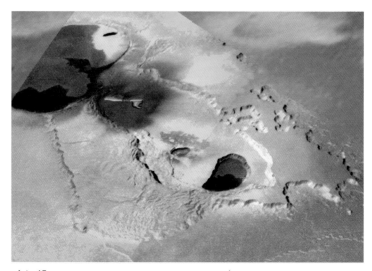

plate 15

A spectacular 300 km wide view of part of Io, a satellite of Jupiter, seen in almost natural colour. A large volcanic caldera occupies the upper left and centre of the view. Dark patches are fresh lava. By good fortune, incandescent magma was erupting from a fissure on the caldera floor and supplying an active lava flow when this rare close-up image was obtained by NASA's Galileo probe in 1999.

Deposits of fine-grained calcium carbonate derived from the shells of marine plankton cover much of the deep ocean floor in places far from land where there is little other sediment supply. It is thought the removal of carbon dioxide into limestone through the action of marine plankton is a natural mechanism that maintains the concentration of carbon dioxide in the atmosphere at an equitable level. This may eventually help to soak up much of the carbon dioxide released into the atmosphere through burning fossil fuels, and so act as a natural brake on mankind's additions to the **greenhouse effect** in which carbon dioxide reduces the Earth's ability to radiate heat to space.

Tiny shells dissolve slowly as they sink, especially in cold waters, and in the deepest and coldest regions deep ocean sediments tend to be dominated by the microscopic shells of planktonic plants called radiolaria that secrete silica rather then calcium carbonate, because no carbonate debris survives there.

In some environments, limestone can form as a ready-made hard rock rather than an unconsolidated deposit. This occurs where the sea floor is encrusted with calcium carbonate-secreting sedentary organisms. Today, the most important such encrusting organisms are corals. These are animals but have photosynthetic algae living symbiotically within them, so they require clear shallow water with sufficient sunlight. Delicate branching or fan-like varieties grow in quiet conditions, and more robust forms colonize rougher settings.

Coral reefs are familiar to most people through marine wildlife television documentaries. If the substrate they are built on is subsiding, they can build upwards so as to keep pace and thereby maintain the living part of the reef close to sea level. This was realized by Charles Darwin in his explanation for coral

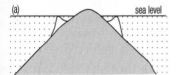 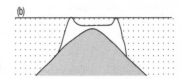

figure 9.13 Cross-section showing two stages in the formation of a coral atoll by subsidence. (a) An extinct volcano is fringed by a reef. (b) The island sinks isostatically (because of the load exerted on the lithosphere by the weight of the island), but the reef builds up to keep pace with the subsidence.

atolls, which are rings of reef made of robust, wave-resistant coral sheltering a shallow lagoon that hosts more delicate corals (Figure 9.13). Drilling has shown that the volcanic basement to some atolls has sunk to more than 1000 metres below sea level. Mururoa atoll in the Pacific, which achieved notoriety in the mid-1990s as the site of French underground nuclear testing, is of this type.

Apart from forming atolls, coral reefs shelter many tropical shorelines in places devoid of muddy river sediment. The biggest and most famous of these is the 2000-kilometre-long Great Barrier Reef off the coast of north-east Australia.

Along some carbonate-rich coastlines, agitation by waves causes a rolling motion that can cause near spherical millimetre-sized grains of calcium carbonate to form, by concentric growth around a tiny fragment or silt particle. These grains are called ooids, and they can accumulate into deposits forming rocks called oolites, or oolitic limestone.

Turning sediment into rock

It has been possible to give only a general account of the more distinctive sedimentary environments. We must finish by considering what it takes to turn a deposit of sediment into a rock.

The main requirement is that it must remain buried, and not be eroded away. It is easy to forget this obvious point when looking at a sedimentary rock in the field. A three-centimetre-thick layer of ripple cross-bedded sand may have taken less than a day to accumulate, but it may be the sole survivor of a million such layers that were successively deposited and then stripped away, perhaps with every turn of the tide. Such is the fate of most sediment deposited near shorelines, in rivers and in desert dunes, and it is only in deep marine conditions that we normally find continuous (but very slow) deposition without erosional breaks.

Once buried, processes begin that can turn soft sediment into hard sedimentary rock. Only reef limestones and some evaporites are formed in an initially consolidated fashion; other sediments begin as loose, unconsolidated deposits. However, with burial, and over time, various chemical and physical changes take place, collectively described as **diagenesis**.

The dominant physical process is compaction, caused by the weight of later sediment deposited on top. Clay-rich muds may begin with solids making up only about 20 per cent of the volume, the remaining 80 per cent being spaces between clay flakes filled by water. The deposit is said to have a **porosity** of 80 per cent. By the time such a deposit is buried by a kilometre or so of sediment, most of this pore water has been squeezed out, and the porosity is reduced to less than ten per cent (Figure 9.14). A one-metre layer of compacted mudrock may thus represent several metres of uncompacted mud.

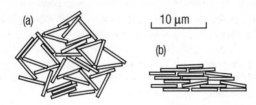

figure 9.14 Magnified cross-sections showing compaction of mudrock: (a) loosely packed clay flakes oriented randomly soon after deposition; (b) clay flakes rotated during compaction so they lie horizontally. In this state the rock can be split readily along closely spaced horizontal planes and is called a shale.

Sands and silts are made of more equidimensional clasts than clay, so the potential for compaction and loss of porosity is less (Figure 9.15). Compaction is achieved by a combination of rotation, bending and fracturing of grains. A sandy sediment may begin with up to 40 per cent porosity and end up with only ten per cent porosity at a depth of about a kilometre. At greater depths, the points of grains resting on others may dissolve away through a process called pressure solution, reducing porosity still further. Similar compaction processes affect carbonate sediments.

Most of the chemical changes during diagenesis are controlled by the nature of the pore waters. Silica that has been dissolved by pressure solution may come out of solution to fill the remaining pore spaces with silica **cement**. This can turn a loose sandy deposit into a hard rock (sandstone) in which the grains are as thoroughly stuck together as crystals in an igneous or metamorphic rock. Pressure solution in limestones results in pore waters that are saturated in calcium carbonate, so calcite

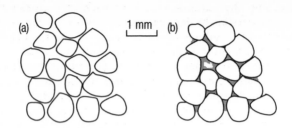

figure 9.15 Magnified cross-sections showing compaction of sandstone. (a) Grains soon after deposition. Each grain touches its neighbours, but only at one point, which rarely lies exactly in the plane of the cross-section. (b) The same grains after compaction. There has been some rotation of grains, but most of the compaction has been achieved by pressure solution, in which material has been dissolved at pressure-points where grains touch. Dissolved material has been precipitated nearby, so that most of the remaining pore space is filled by cement

cements are almost universal within limestones. Calcite cements may also form in the pore spaces of other rock types, either from solution of minor amounts of calcium carbonate originally in the deposit, or from calcium carbonate carried in from elsewhere by migrating pore waters.

There are some mineralogical changes that may take place too. One of the earliest of these affects limestones. In many shells, the calcium carbonate takes the form of aragonite, a polymorph of calcite. Fragmental debris created from the abrasion of shells is therefore mostly aragonite, but this converts to calcite during diagenesis, so hard rocks made of aragonite are unknown. Sometimes the diagenesis of limestones from originally lime-rich muds involves so much recrystallization that the resulting rock consists largely of interlocking crystals of calcite. Its texture may resemble an igneous rock rather than a fragmental sediment, although its composition and the common presence of fossils makes misidentification unlikely.

Other diagenetic changes include the driving off, as a result of pressure and mild heat, of water that was originally bound up in clay minerals. New clay minerals, with less water in their formulae, take their place. This sort of change is on the verge of metamorphism, and it is really just a matter of choice what to describe as extreme diagenesis and what to think of as low-grade metamorphism.

We have now seen how sedimentary rocks are created from components derived by weathering and erosion. Sedimentary rocks are not a dead-end in the rock cycle. Most ocean floor sediments eventually get dragged into trenches at destructive plate margins, where they are metamorphosed or even melted. Sediments in shallow seas or deposited on land may be regionally metamorphosed after burial to sufficient depth, or if they are caught up in a continent–continent collision. Alternatively they may be raised up to be exposed to the forces of weathering and erosion. The remaining links in the rock cycle are covered in the next chapter, where you will see how rocks behave when subjected to stresses that deform them.

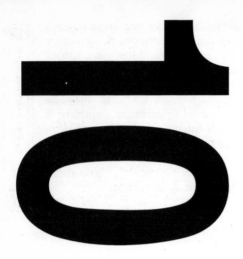

10

deformation of rocks

In this chapter you will learn:

- about the ways in which rock can become deformed, and the processes responsible
- how solid rock can fracture when it is displaced by fault movement, and how in other circumstances it can be bent into structures known as folds
- how a deformed sequence of rocks can become buried by younger undeformed sediments.

Parts of the Earth's crust are from time to time subjected to forces that tend to extend or compress them. These forces may be regional in extent, such as those associated with plate tectonics, or may operate more locally, perhaps caused by the extra load on the crust produced by a thick accumulation of sediment or the growth of a volcano.

When rock is relatively warm, it is able to deform in a **ductile** fashion, that is to say it can be squeezed or stretched without fracturing. This occurs by rotation, deformation and recrystallization of grains or crystals within the rock, and happens most readily if the deforming forces build up slowly. Ductile deformation is thus characteristic of deformation at deep crustal levels, where the temperature is higher. At shallower levels, rocks tend to fracture, especially if forces build up quickly. This is how faults and joints form, and is described as **brittle** deformation.

The effects of deformation are most readily seen in sedimentary rocks, because their inherently bedded nature provides ready-made markers. For example, an individual bed of limestone, sandstone or mudrock may be traceable over tens or even hundreds of kilometres. If its base and top were virtually horizontal when deposited (provided we are sure we are looking at the real top and bottom boundaries of the bed, rather than cross-bedding within it), and if the bed is no longer horizontal when we see it then it must have been tilted. Similarly, if the bed is no longer flat but curved, then we can deduce that it has been folded. Moreover, some beds within a sedimentary sequence are usually distinctive, and enable us to match displacements across faults.

Different types of steep fault

We will start by looking at faulting, which is brittle displacement of rock. There are essentially three types of fault movement, all illustrated in Figure 10.1. The rock all around has been made invisible to enable us to see what happens to a specific block of terrain. The sequence here consists of an undeformed succession of horizontal sedimentary beds cut by a dyke of igneous rock, which will act as a convenient vertical marker. For our purposes, the sedimentary beds could instead equally well be lava flows, layers of pyroclastic rocks or a series of horizontal sills; the main thing is that there is a set of individually recognizable horizontal layers. A useful term that encompasses all kinds of layers, not just sedimentary beds, is **strata**.

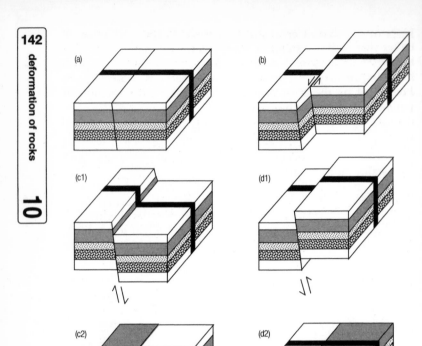

figure 10.1 Cross-sectional block diagrams illustrating three types of fault, each starting from the situation in (a): (b) shows the area after transcurrent movement across the fault plane; (c) shows what would happen if, instead, the block on the right went down relative to the block on the left (c1 without erosion, c2 after erosion); (d) shows what would happen if the block on the right went up instead of down. Note that the fault plane is not quite vertical. We can therefore distinguish case (c) as a normal fault and (d) as a reverse fault. See text for further discussion.

In Figure 10.1(a) the area is shown before any fault motion has taken place, though the position of the future fault is marked. In this example it is a plane that is not quite vertical but is dipping steeply down to the right.

One thing that might happen is that the blocks of terrain on either side of the fault might slip sideways past each other. This is called strike-slip or **transcurrent fault** motion, and is what is happening along most of the San Andreas fault in California. As Figure 10.1(b) shows, if the direction of motion is exactly parallel to the bedding, there is no resulting mismatch of displaced strata either side of the fault. However, the vertical marker, in this case a dyke, has been offset, and the amount of offset can be measured to determine how much movement there has been across the fault plane. If we were to stand on the dyke on one side of the fault, we would see that the other part of the dyke had been offset to the left. We would see a leftwards offset whichever side of the fault we were looking from. The fault illustrated is thus distinguished as a left-lateral (or sinistral) transcurrent fault. If the offset had been in the opposite sense, so that the dyke had been displaced to the right, it would be called a right-lateral (or dextral) transcurrent fault.

Alternatively, the fault plane could separate regions whose relative movement is vertical. When this happens, the offset can be measured by the fact that strata that were originally at different depths have become juxtaposed (it is immaterial whether we regard one side as going up or the other side as going down). However, there will be no evidence of offset displayed by a vertical feature cut by the fault. Figures 10.1 (c) and (d) show alternatives, each starting from the situation shown in Figure 10.1 (a). In Figure 10.1 (c) the terrain on the right has gone down relative to that on the left. Usually, the side that is being relatively uplifted experiences much more erosion, and the result of planing off to a new horizontal surface is shown in Figure 10.1 (c2). As a result, an older bed on the left of the fault is exposed at the surface next to a younger bed on the right of the fault. In Figure 10.1 (d) the fault motion was in the opposite direction, and the younger bed survives at the surface on the left of the fault instead.

Figures 10.1 (c) and (d) are not quite mirror images of each other, because the fault plane is dipping steeply downwards to the right instead of being vertical. In Figure 10.1 (c) the downthrown terrain is on the down-dip side of the fault; this is the commonest kind of fault movement, and is known as a **normal fault**. In Figure 10.1 (d) the downthrown terrain is on the up-dip side of the fault. This is described as a **reverse fault**. In active fault belts, erosion may keep pace with the rate of uplift, so that faults are rarely represented by cliffs. There is certainly never an overhang at the surface of the kind shown in Figure 10.1 (d1), because this would crumble away as soon as it formed. However, sometimes erosion lags behind to the extent that the position of an active normal or reverse fault is indicated by a straight edge to a belt of hills.

If you compare stage (c) in Figure 10.1 with stage (a), you will see that because the fault plane is not quite vertical, the distance from left-to-right has increased slightly. Normal faulting is therefore common in regions of crustal extension. If, however, the fault acts as a reverse fault, ending up at stage (d), the left-to-right distance has been slightly reduced. Reverse faults are thus characteristic of regions of crustal compression. However, these are gross generalizations. In nature most faults are parts of linked systems and it is unwise to read significance into a single fault.

A series of consecutively developed normal faults appeared in Figure 4.6, which showed continental rifting. Figure 10.2 shows a similar situation at an enlarged scale, and defines some terms that you may find used elsewhere to describe faulted areas. Note that the faults in this figure are shown as curved surfaces, becoming less steep at depth, which allows rotational movement between adjacent blocks.

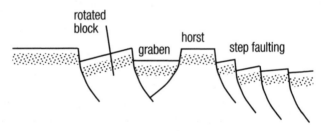

figure 10.2 Cross-section showing some features associated with faults. A graben is a block that has dropped down between two normal faults. A horst is a block that has been left standing high because blocks on either side have dropped down. In nature, the scale of this diagram would typically be hundreds of metres to tens of kilometres across.

Low-angle faults

The fault plane drawn in Figure 10.1 dips steeply, and cuts the bedding at a steep angle. Sometimes fault planes lie at a low angle to bedding, and may even become parallel to it. In this case, where the motion is compressional, the fault is described as a thrust fault, or more commonly just as a **thrust**. Where the motion is extensional, the fault is described as a low-angle normal fault or low-angle extensional fault. Faults that develop parallel to bedding do so because some layers are easier to fracture than others, and thrusts and low-angle normal faults usually lie between beds or within weak beds such as shales, and avoid stronger beds such as well-cemented sandstones and limestones. However, they must approach the surface somehow, and this is usually achieved by the fault plane cutting steeply through each strong bed yet dipping gently everywhere else. In this situation it is more appropriate to write of a fault surface rather than a fault plane, because it is certainly not a smooth plane.

Figure 10.3 is a rather complicated block diagram, but it serves to illustrate many of the most important features of low-angle faults. A thrust is illustrated, but the same geometrical concepts apply to low-angle normal faults. Stage (a) shows the situation before displacement, with the position of the incipient fault surface marked. The rock below the fault is described as being in the **footwall**, whereas the rock above belongs to the **hangingwall**. Thrust movement is going to occur from left to right parallel to the side of the block facing us. Looking at this face first, we can see that the fault surface runs between beds for most of its length (perhaps there is a thin, weak layer of mudrock to lubricate movement between each of the major beds drawn), but cuts steeply upwards at two places. The steep bits are called footwall ramps, and the flat bits are called footwall flats. The fault does not reach the ground surface on this face, but presumably we would see it do so if the diagram were extended to the right. The right-hand face of the block reveals the fault surface at the base of the topmost bed close to the near corner, but the fault surface turns upwards to reach the ground about a third of the way along the face. If we trace the fault onto the top (ground) surface, we see that it becomes a fault line running parallel to the direction of thrust movement. This sort of boundary is referred to as a sidewall ramp, and along this part of its length the fault will behave as a transcurrent fault. The fault at the surface makes a right-angle turn where it becomes a thrust again, then turns again to become another length of transcurrent fault.

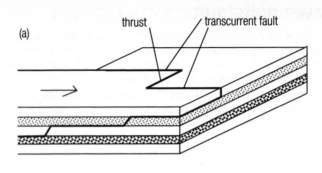

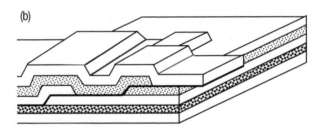

figure 10.3 Cross-sectional block diagrams illustrating motion of a thrust sheet. In nature, the scale of this could be anything from a metre to tens of kilometres across. See text for discussion.

To put it more concisely, the sheet of rock that is about to be transported by a thrust (we would call it a thrust sheet) is bounded on its lower surface by a thrust fault and on its sides by transcurrent faults.

The situation after thrust movement is shown in stage (b). Looking at the near face of the block we can see that there are two places where a bed has been repeated by being thrust over itself. Also, because the fault surface is not a plane but has steps in it, the thrust sheet has had to deform, in fact it has folded in a ductile fashion. This folding occurs only in the thrust sheet, and it does not affect the units in the footwall. In nature, the up-folded areas would probably become eroded away, but the changing dips of the beds exposed at the surface would demonstrate that folding had taken place, and would enable us to deduce the locations of underlying footwall ramps.

Thrust belts and collision zones

You will look at folding in more detail shortly, but first there is another important aspect of thrusts to consider. You saw in Figure 10.3 that the footwall rocks below the thrust surface do not fold like those above. However, very often as deformation proceeds, a new thrust develops below the first. What happens is that the active thrust surface jumps forward by slicing off ramps, as illustrated in Figure 10.4.

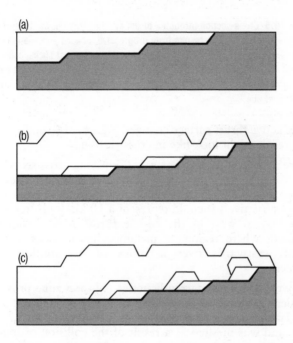

figure 10.4 Progressive development of a thrust system, seen in cross-section. Undeformed rock below the lowest thrust is shaded. Bedding is not shown, but can be assumed to be horizontal, except where folded as a result of thrusting. In each case, the thrust that is just about to become active is shown by a heavy line. Inactive lengths of thrust are shown by finer lines. Note that these become folded when they are transported over footwall ramps.

In this way we develop a series of blocks (or 'horses') bounded above and below by thrusts, called a duplex complex. Rocks in the hangingwall that were folded during motion of the original thrust sheet are liable to be folded again each time the geometry

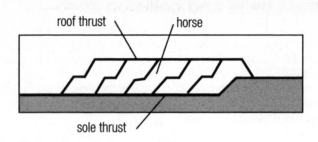

figure 10.5 Cross-section through a thrust system. This shows the sort of structural complexity that results from progressive forward propagation of the active thrust, as in Figure 10.4. A series of imbricated thrust slices, called 'horses', is produced, each horse being bounded by thrusts above and below. In this example, thrust motion is from left to right, so the first horse to form was the leftmost one.

of the active thrust surface changes. Figure 10.5 shows what the final result might look like in cross-section.

Figure 10.5 could represent a 100-km-long cross-section from west to east under the eastern Canadian Rockies. These are a collisional mountain belt with thrust movement from west to east. Alternatively it could represent a 10-km-long cross-section from east to west of the Moine thrust zone in north-west Scotland, where regionally metamorphosed rocks have been thrust westwards over a series of 500-million-year-old sediments.

The world's biggest currently active collision zone provides an excellent example of how different kinds of fault are related. The Tibetan plateau has an average altitude of 5 km above sea level, and was uplifted as a result of the collision of India into central Asia, which began about 35–50 million years ago. The plateau is bounded to the south by the Himalayas, which is a mountain belt that has developed by means of a series of southwards-directed thrusts. The plateau itself is crossed by many grabens (faulted valleys) delineated by normal faults running north to south, which at first sight seems odd in a region that ought to be experiencing north–south compression. The explanation becomes clear when it is recognized that the plateau is bounded by and crossed by numerous transcurrent faults. Some run from south-west to north-east and are left-lateral faults, others run from north-west to south-east and are right-lateral. Tibet is buttressed by mountains on all sides except

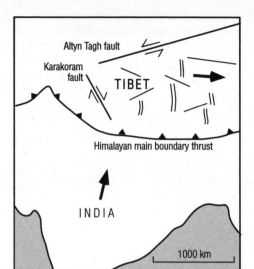

figure 10.6 Sketch map showing how the Tibetan plateau is bounded and crossed by transcurrent faults, and has north–south normal fault systems within it. Tibet is being squeezed out towards the east. The fault systems in Tibet are so large that they were not appreciated until seen from space.

the east, and is being squeezed out eastwards by motion across these faults. The north–south normal faults are just a manifestation of Tibet's east–west extension that accompanies its north–south shortening (Figure 10.6).

So, to really understand the tectonic history of an area, the whole fault system needs to be mapped out. Sometimes it can be deduced that fault motion has reversed one or more times during a prolonged period of deformation, or that faults that seem related were in fact active at quite different times.

Brittle faults versus shear zones

The best-known ancient faults are those that were active at relatively shallow depths in the crust, a few kilometres or so, because these are the ones that are most likely to end up exposed at the surface by erosion. The rock within a few centimetres or metres of such a fault is often intensely fractured, and is described as a fault **breccia**. Faults therefore tend to be zones of weakness that erode away readily. Because of the fracturing associated with them, faults may act as pathways for groundwater, and can be important sources of underground water in arid regions, especially where they occur in hard igneous or metamorphic rock that is otherwise impermeable.

Low-angle faults can be mapped below ground because they will reflect back seismic waves generated by explosions deliberately set off for the purpose. It is believed that below a depth of 10–20 km, faults become ductile features. Displacement is accommodated across a shear zone, which is tens or hundreds of metres wide for moderate-sized faults.

Folds

Folding is another kind of ductile deformation, and it can occur at shallow depths. We have already seen that **folds** are an inevitable side-effect of thrusting. The Appalachian Mountains are part of a folded and thrust mountain belt that stretches from Alabama to Newfoundland and is the result of a continent–continent collision between North America and Africa that took place about 300 million years ago.

Traditionally, folds have been regarded as features produced when the crust experiences compression, however, they can also form in hangingwalls in extensional settings, as a consequence of footwall ramps on low-angle extensional faults.

Figure 10.7 shows a cross-section through folded rocks, to introduce some terms commonly used to describe folds. An **anticline** is an arch-like structure, whereas a **syncline** is U-shaped in cross-section. The axis or hinge of a fold is defined as where the curvature is greatest, whereas in the limbs on either side of the axis curvature is less.

Shapes and styles of folding

Folds can take a whole variety of forms, depending on the rate of deformation, the depth and the mechanical properties of each stratum. The strata in Figure 10.7 are folded in continuous curves, in contrast to the folds portrayed in Figure 10.3 where the changes in dip are sharp. In Figure 10.7 there is a tendency for the strata to get slightly thicker near the fold axis: this is

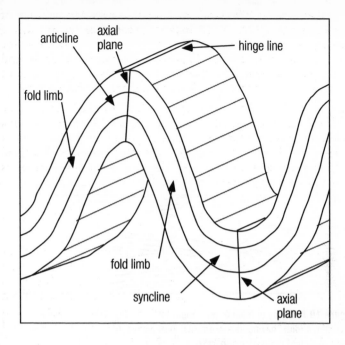

figure 10.7 A perspective cross-section through some folded strata. This could be on any scale from centimetres to kilometres across.

accomplished by ductile deformation within each stratum, manifested by distortion of features within it. In extreme cases, the degree of distortion and accompanying rotation and recrystallization can lead to the development of cleavage, especially in mudrocks, with cleavage planes being roughly parallel to the axis of the fold. To find cleavage in folded rocks should come as no surprise to you, because you met cleavage as an essential element of low-grade regional metamorphism in Chapter 07, and by their very nature most regionally metamorphosed rocks have been deformed.

The axial planes of the folds in Figure 10.7 are very steep, and the hinge lines are horizontal. It is not always so. Axial planes can sometimes have very gentle dips, hinge lines can plunge downwards at any angle, and opposite limbs of a fold can diverge at a larger or smaller angle (Figure 10.8). In highly deformed regions, fold limbs are often parallel to one another, and in these cases the folds are described as isoclinal. If the axial

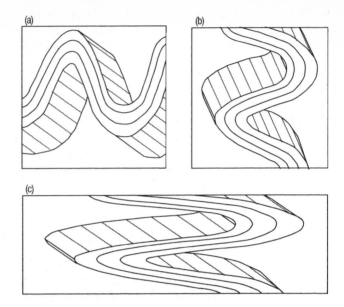

figure 10.8 Styles of folding: (a) upright folds, having steep axial planes; (b) recumbent folds, having near-horizontal axial planes; (c) isoclinal recumbent folds, with fold limbs almost parallel.

figure 10.9 Recumbent folds in thinly bedded siltstones on the north Devon coast, England.

plane dips at less than ten degrees the fold is described as recumbent (Figure 10.9). In very multiply deformed areas, younger folds may distort older folds, resulting in a complex pattern (Plate 10). Folds may even be turned upside down, in which case the anticlines look like synclines, and vice versa. In such a situation the right way up has to be determined by inspection of small-scale structures such as cross-bedding.

Folds manifest themselves on the landscape in a variety of ways. It is rare for the crests of hills or mountain ranges to correspond with the axes of anticlines. It is the more resistant strata that form the high ground, as shown in Figure 10.10.

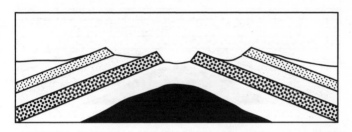

figure 10.10 Cross-section through an eroded anticline, looking along its hinge line, showing how strata that are more resistant to erosion form ridges. Gentle slopes following the surface of a stratum are described as 'dip slopes'. Opposite-facing steeper slopes where a stratum has been cut through by erosion are called 'scarp slopes'.

Unconformities

Sometimes tilted or folded strata that have experienced erosion are then covered by younger deposits. These will share neither the dip nor the deformation history of the strata they bury. The junction between the two series of rock, which may represent a gap of millions or even hundreds of millions of years, is called an **unconformity**.

An angular unconformity, as in Figure 10.11, is usually easy to spot because of a difference in the amount or direction of dip between the strata below and above. However, it can be very difficult to recognize unconformities where there is no difference in dip. These occur where the upper strata of an unfolded, untilted sequence have been eroded away in a planar fashion and then covered by younger strata, or simply where deposition

(a)

(b)

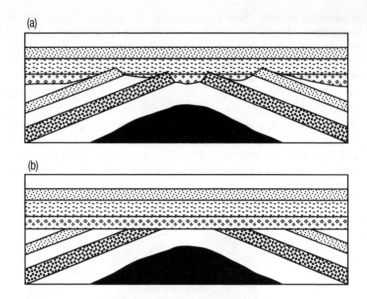

figure 10.11 Cross-sections showing younger strata deposited unconformably on the eroded anticline seen in Figure 10.10: (a) no further erosion before deposition, so that old topography is preserved in buried form below the younger strata; (b) the older rocks planed flat before deposition recommenced. In both cases the junction between the older (folded) and younger sequences is described as an angular unconformity, because of the difference in dip between the strata above and below.

was interrupted by a pause during which no sediment accumulated.

Although, technically, any subsequently buried erosional surface can be called an unconformity, the term is not usually applied if the presumed time gap is geologically short, the amount of erosion slight, and the change in environment trivial. For example, the erosional bases of the river and delta channels in Figures 9.8 and 9.11 would not normally be classed as unconformities.

However, in the case of angular unconformities, common sense and logical thinking can tell you that a very considerable time must have elapsed between the deposition of even the youngest rocks below an angular unconformity and the oldest rocks above it (time is needed to first deform and then erode the sequence below, before deposition recommences). This realization, first articulated by the Scotsman James Hutton (1726–97), led geologists in the late eighteenth century to challenge the widely-accepted young age of the Earth (a few thousand years) implied by a literal interpretation of the Bible.

One way to decide how long a time interval is represented by an unconformity is to determine the difference between the ages of the beds immediately above and below. In many situations, the most precise dating comes not from radiometric techniques such as those you met in Chapter 02, but by looking for differences in any fossils. You will examine fossils and their value in stratigraphic dating in Chapter 12, but first we will consider the Earth's physical resources.

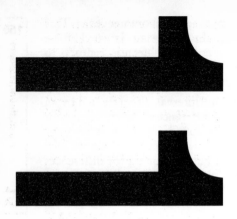

11

physical resources

In this chapter you will learn:
- about fuels and other useful materials that can be extracted from the ground
- how some of them originate, and how they can be found
- about the implications of extracting them much faster than they are renewed by natural geological processes.

We rely on geology for all our raw materials except things that grow, and for almost all our energy except solar power. The geological processes of winds, waves and tides can be used to generate power, but the vast majority of power to run our industrial society comes from burning **fossil fuels**, specifically coal, oil and natural gas – all of which release carbon dioxide into the atmosphere when burned, and so contribute to global warming via the greenhouse effect. Because we are using up these fuels at rates vastly in excess of the rates at which geological processes can form them, they are classed as **non-renewable resources**, along with the radioactive fuels used for nuclear power. Wind, wave and tidal power, and (to an extent) geothermal power generated from the heat escaping from shallow magma bodies, do not outstrip the rate of supply, and are referred to as renewable resources.

In addition to fossil fuels, non-renewable resources include such commodities as the ores from which metals are extracted, the clay used to make bricks and porcelain, the limestone used to make cement, and rock in general used to make road surfaces.

If we continue to use up raw materials at the present rate, some things will run out sooner than others. For example, there is no shortage of limestone that could be used to make cement. We might not be willing to accept the environmental impact of quarrying away beautiful limestone hills, but that is a separate issue. On the other hand, in 2007 silver was being used up at a rate sufficient to exhaust the world's supply of that metal within 13 years.

The rates at which we are using up many raw materials, and the inadequacy of recycling, are causes for concern, but there is no need to panic. The world will not actually run out of silver in 13 years. More silver deposits may be discovered, or else silver will have to be extracted from deposits that are currently uneconomic to exploit. This will push the price of silver up, and encourage both recycling and the use of substitutes.

Reserves and resources

This brings us to the important distinction between **resources** and **reserves**, two terms that are often misused. Their relationship is summarized in Figure 11.1. Strictly speaking, 'reserves' means only those raw materials than can be extracted profitably and legally under existing conditions. On the other

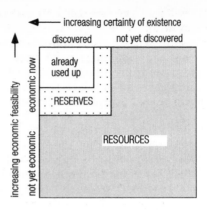

figure 11.1 The relationship between physical resources and reserves.

hand, 'resources' should be used only in a wider sense, to describe the total amount of the raw material that could be extracted if the price were high enough, or legislation allowed, or technological advances made it feasible. Because exploration is incomplete, estimates of resources may be wrong.

The earlier statement that silver would run out in 13 years is thus rather inadequate. It should be rephrased to state that the current *reserves* of silver were due to be exhausted in 13 years. In practice, as these reserves are used up by extraction they will be added to by decisions to extract silver from deposits that were formerly regarded as not worth exploiting, or by the discovery of new economic deposits. Thirteen years is thus a fairly comfortable cushion, so long as new discoveries continue at a rate sufficient to keep pace with extraction. Even so, every non-renewable resource is liable to be exhausted eventually, so recycling and the use of substitutes will inevitably become increasingly important. Thus, we should extract non-renewable resources with care and sensitivity, and consume them wisely.

Global reserves of uranium (used as a fuel in nuclear reactors) are about 70 years' supply, and the untapped resources are vast. Nuclear power is attractive because of the almost infinite nature of its fuel supply and because it does not emit carbon dioxide and thereby contribute to global warming. However, a significant downside is that spent nuclear fuel is highly radioactive and needs to be stored or buried securely.

In this chapter we will first look at fossil fuels, and then at other non-renewable resources such as **ore** minerals and bulk materials. We will see how their origins can be explained by the processes that we encountered earlier in this book.

Origins of oil and gas

Of all the fossil fuels, oil is the one that tends to make the headlines. It has become so important to the global economy that wars have been fought over it. Oil and natural gas have the same origin, and are often found in association. Moreover, they are both **hydrocarbons,** so we will consider them together. They are also the fuels whose burning, along with coal, is responsible for most of the human-induced increase in atmospheric carbon dioxide, leading to global warming by means of the greenhouse effect.

The raw material for hydrocarbons is dead organic matter, chiefly the remains of micro-organisms and land plants. For enough of this material to collect, the environment must have a large supply of dead organic matter coupled with a lack of oxygen to stop it decaying to water and carbon dioxide. These conditions are met in areas of the sea floor where the surface waters have a high rate of production (and mortality) of planktonic plants, which upon death sink to the bottom to be buried in fine-grained, shaley sediment. More rarely, the floors of lakes provide a suitable environment.

Once an organic-rich shale has accumulated, it is a potential **source rock** for hydrocarbons. For this potential to be realized, burial must occur, to allow heat, pressure and chemical conditions to cause the necessary diagenetic changes in the organic matter. The conversion from dead organic matter to a hydrocarbon is described as maturation.

By the time burial has reached a depth of 1 km, the temperature is likely to be about 50°C. Under these conditions, the organic matter will be cooked into complex molecules called kerogens, which consist of carbon, hydrogen and oxygen. With deeper burial to about 6 km and temperatures up to about 180°C the kerogens break down to chains of carbon molecules bonded with hydrogen, which is petroleum. The general formula is C_nH_{2n+2}, where C is carbon and H is hydrogen. The chains get shorter according to the temperature of maturation. Where n is 16 or more the product is a waxy solid, but shorter chains are liquid.

At slightly higher temperatures the chains become really short, producing C_4H_{10} (butane), C_3H_8 (propane), C_2H_6 (ethane) and CH_4 (methane), which collectively are known as natural gas. If the temperature rises much above 200°C even methane breaks down, driving off the hydrogen and leaving the carbon behind as graphite. Thus hydrocarbons can form and survive only if conditions stay within a fairly narrow band. There is actually a time element too, and hydrocarbons can form and break down at lower temperatures given sufficient time (Figure 11.2).

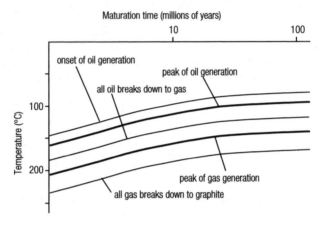

figure 11.2 Production of hydrocarbons by maturation within a suitable source rock, showing how this depends on temperature and time.

Oil and gas reservoirs

It is one thing for hydrocarbons to form and to survive within a rock. It is quite another matter to be able to extract them. If the source rock becomes exposed at the surface, then it can simply be dug up to get at the oil within it. This is the only easy way of extracting non-liquid hydrocarbons or kerogens, but the environmental costs of this sort of quarrying are so high that most oil shales, as they are known, have not yet been exploited.

Liquid hydrocarbons are almost always extracted by drilling. You might think that oil and gas wells are made by drilling into source rocks. This is not the case, because although shales are porous and can contain ten per cent or so by volume of hydrocarbons, they are not permeable. That is to say the pore-spaces are not sufficiently well connected to allow fluids to flow through the rock at the rate required to feed a well.

Most hydrocarbons are extracted not from their original source rock but from a **reservoir rock**, into which they have migrated over geological time. A reservoir rock must be both porous and permeable. Sandstones usually fit the bill, unless most of the pore-space has been filled by cement, and these are the most common reservoir rocks for hydrocarbons. A rubbly limestone, formed in a reef, may make a good reservoir rock too, especially if ground water has previously dissolved some of the rock, thereby increasing its porosity and **permeability**.

In order for hydrocarbons to be held within a reservoir rock, there must be a nearby source rock from which the hydrocarbons can have migrated, and the reservoir rock must have remained below the temperature at which the hydrocarbons break down. Once they have found their way into a potential reservoir rock, oil and gas tend to be displaced upwards by pore water, which is denser than oil. A final condition therefore, is that the hydrocarbons must somehow be trapped within the reservoir rock, otherwise they would have long since seeped out to the surface.

There are essentially two kinds of oil trap. Structural traps are when the reservoir rock is folded or is next to a fault in such a way that the hydrocarbon-bearing part is bounded by impermeable rocks, so the hydrocarbons cannot escape (Figure 11.3a and b). Stratigraphic traps occur below an unconformity (Figure 11.3c).

Once the appropriate part of a hydrocarbon reservoir has been drilled into, oil or gas may be driven to the surface merely by pressure. Usually about 30–40 per cent of the oil may be recovered this way. Most of the remainder can be extracted by displacing it by pumping water or gas down a suitably positioned borehole.

Exploration for oil and gas

Oil wells are usually 1–5 km deep. It is very expensive to drill such deep holes, so drilling is not the major technique used in exploring for hydrocarbons. Sometimes, interpretation of the surface geology may suggest that an oil trap exists at an appropriate depth, but often these structures are too deeply

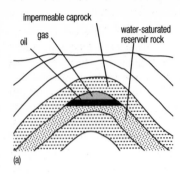

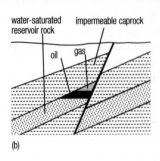

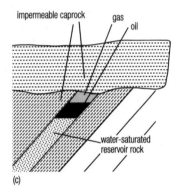

figure 11.3 Cross-sections to show oil traps: (a) a structural trap in the form of an anticline or a dome; (b) a structural trap against a fault; (c) a stratigraphic trap below an angular unconformity. In all three cases, impermeable rock prevents hydrocarbons from escaping through the sides and top of the trap. The areas labelled 'oil' and 'gas' show where the reservoir rock is saturated with these.

buried to be apparent at the surface. In such a situation, initial exploration relies on geophysical techniques such as mapping anomalies in the local magnetic and gravity fields.

When an area appears promising, it is worth making a seismic study. This involves using explosions to generate seismic waves, which are picked up by an array of sensors. Seismic waves generated this way are not as powerful as those from a major earthquake, but they have the advantage of providing a sharp signal, generated from a chosen point at a known time. By interpreting the reflected seismic signals from successive rock layers, a picture of the sub-surface can be built up.

Surprisingly, seismic surveys are most easily done under water. A ship trails a kilometre-long array of sensors to record pressure variations in the water, and periodically generates an underwater explosion using a compressed air gun. Part of the energy is reflected back by the sea floor (this is the principle of echo-sounding to determine water depth), but the rest continues into the sea-floor sediment, with a portion of it being reflected back at each interface between layers.

When an anticlinal or domal structure, or some of other kind of potential hydrocarbon trap, is revealed by a seismic survey then it is worth drilling an exploration hole. In addition to attempting to reach the hydrocarbons, much can be learned by studying rock chippings deliberately flushed out of the borehole, and by lowering various electronic, magnetic and radioactive sensors down the hole to record 'wireline logs' giving estimates of the density, porosity, composition and even the dip of bedding for each unit that is passed through. Only when a successful exploration well has been drilled and the extent and capacity of the reservoir have been ascertained is it worth sinking production wells. These may be drilled at any angle, and, especially for offshore production, a production platform may extract hydrocarbons from a three-dimensional fan of production wells that penetrate the reservoir over an extensive area.

The origin of coal

The other great fossil fuel is coal. Like the hydrocarbons, coal derives from dead organic matter, but in this case the source is mainly land plants. Several criteria must be met for coal to form. First, there must be abundant land plants living (and dying) nearby. Coal is unknown in rocks older than about 425 million years, which is when the first land plants appeared. It is particularly abundant in sequences formed at certain times between about 350 and 50 million years ago when climate, the arrangements of continents, and the evolutionary state of land plants combined to favour its formation (Plate 11).

The second requirement, as for hydrocarbons, is that the dead plant material should accumulate in an environment lacking oxygen ('anoxic') to prevent oxidation. Swamps meet these first two criteria very well, having abundant vegetation and a stagnant water table coinciding with the sediment surface. The organic-rich soil here is described as peat.

The next criterion for an economic coal deposit is that there should be rapid subsidence to allow a sufficient thickness to accumulate. Ten metres of peat compacts down to only one metre of coal. The final requirement is that the deposit should be buried to hundreds of metres to enable the peat to be turned into coal. All four requirements are met in delta environments, where, as we saw in Chapter 09, a swampy delta top where peat can accumulate is continuously subsiding under the weight of sediment. If the main distributary channel of the delta switches, the old delta top may be buried by marine or delta-front sediments (Figure 9.11). The major coalfields of Europe, Asia and North America formed in this way about 300 million years ago, in the latter half of the period of geologic time called the Carboniferous (sometimes distinguished by North American geologists by the name 'Pennsylvanian').

The quality or 'rank' of coal depends principally on how deeply it was buried during diagenesis. Under anoxic conditions, dead plant material is broken down by bacterial action to a woody residue called lignin. With progressive burial, volatiles such as water and methane are driven off, the deposit becomes compacted, and the amount of energy that can be released by burning a given weight of fuel increases. Peat has less than 50 per cent carbon, but the highest rank of coal, anthracite, is over 90 per cent carbon, and (weight-for-weight) contains about three times as much energy. Apart from the rank of coal, other factors determining its usefulness include impurities within it. Clay minerals leave a solid residue (ash) after combustion, iron sulphide produces the acid gas sulphur dioxide during combustion, and sodium chloride corrodes boilers.

Mining techniques

Coal can be extracted either by deep mining or by strip mining. The latter (Figure 11.4) is economic for shallow deposits, but becomes prohibitively expensive (and more environmentally damaging) as the thickness of overburden that must be removed to get at the coal increases. Strip mining is the favoured technique in the great US coalfields of Pennsylvania.

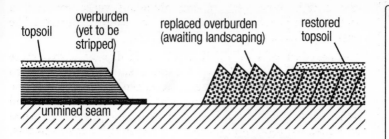

figure 11.4 Area strip mining of a shallow coal seam, showing how the overburden that has to be removed to work the active face can be dumped behind. This obviates the problem of how to dispose of it and (provided the topsoil has been kept separately) the landscape can be restored to something like its original form. The same technique was historically used to mine shallow ironstone seams in England.

In deep mining, a shaft is sunk to the level of a coal seam that is thick enough (typically 0.5–3 m) to be worth exploiting. Coal was formerly removed by men with pickaxes, but nowadays mechanical cutters are used. These make it cheaper to exploit thick seams, but have made it uneconomic to work thin coals underground. There are essentially two approaches to mining of coal, or any other sheet-like deposit (such as evaporite minerals). Each is dictated by the need to prevent the roof collapsing on the miners and equipment. The simpler and more traditional method is pillar-and-stall working, in which only about 50–85 per cent of the deposit is removed, leaving the rest as pillars to support the roof (Figure 11.5a). It is a matter of fine judgement how thick and how closely spaced the pillars need to be if they are to bear the weight of the roof; the thicker and more numerous the pillars, the smaller the fraction of the coal you can extract. The alternative method is longwall mining in which an underground face is created that may be several hundreds of meters long. A cutter operates along the seam beneath a roof supported by hydraulic jacks, and these are slid forward as the cutter passes, allowing the roof to collapse behind it (Figure 11.5b). Access is maintained by a tunnel joining each end of the face to the mine haulage roadways. As the face advances, the ground surface subsides, the effects of subsidence being more extreme if the mine is shallow. However, this method allows the whole of a seam to be extracted.

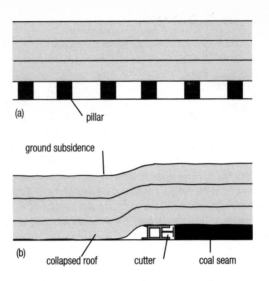

figure 11.5 Techniques for mining a coal seam. (a) Pillar-and-stall working. This works only if the roof is a strong rock type such as sandstone or limestone. If the roof is weak shale it is prone to collapse, unless the pillars are very closely spaced. (b) Longwall mining, seen looking along the face that is being worked. The cutter is moving into the page (i.e. away from us), and the roof is allowed to collapse behind it. When it reaches the end of the face, the cutter will be moved to the right and then begin to work its way back towards us.

Longwall mining is the main method of extracting coal in Britain's few remaining coal mines. In North America, pillar-and-stall working is favoured, although pillars are usually removed when the rest of the seam has become exhausted, allowing the roof to collapse in a controlled manner like in longwall mining. China accounts for a third of the world's coal production, but such is its rate of industrial growth (demanding coal-fired electric power stations) that in 2007 it became a net importer of coal. Chinese mining methods are hard to ascertain, but risks are clearly taken because even the official figures for 2006 admit to nearly 5000 deaths among miners, 100 times as many as in the USA where production was half that of China.

Ores and their origins

Metals are extracted from ores. Ore is a term referring to any rock containing a sufficient concentration of a metal for it to be extracted economically. The concentration of metal (the ore's 'grade') necessary to make it economic depends on the value of the metal and the cost of extracting it. Iron is abundant and cheap, and at present it is not worth extracting from ores containing less than 50 per cent iron. It uses a lot of energy to isolate aluminium from its ores, so it is not usually economic to work ores less than 30 per cent aluminium. Lead is worked from ores whose grade is as low as five per cent, copper from ores around one per cent grade or less, silver from ores of 0.01 per cent (100 parts per million) grade, gold from ores of one to ten parts per million grade, and platinum from ores as low as 0.1 parts per million grade. When several metals occur in the same ore, it can be profitable to work it for the combination of metals, although it would have been uneconomic to extract just one of them.

Most ores result from igneous activity. Sometimes economic concentrations are formed within a magma body when heavy crystals such as chromite (the main ore mineral of chromium, and an important source of platinum and related metals) or nickel, iron and copper sulphides sink to the bottom. A prime example of this type of differentiated igneous ore body is the 200-km-wide Bushveld Complex to the north of Johannesburg, South Africa, which contains enormous reserves of chromium, iron, platinum and gold. One of the most famous, and largest, nickel ore bodies is the 50-km-long Sudbury igneous complex in Ontario, Canada. Interestingly, this is thought to have originated not as the ordinary sort of crustal melt but as a melt formed when a 10-km-diameter asteroid hit the Earth nearly 2 billion years ago.

Other ores form not within an igneous body itself, but in the surrounding rocks. These are a result of hot watery fluids circulating between a cooling intrusion and the surrounding rocks. The process is one of convection, driven by the heat of the igneous body, and is described as hydrothermal ('hot water') circulation. Essentially what happens is that metals are dissolved out of the intrusion by the hot fluids, and deposited nearby, usually as sulphides and sometimes as oxides. The major 'porphyry copper' ore bodies in the Andes are of this type, as are the tin ores in Cornwall, England (where parts of the granite have been altered to kaolinite or china clay by the hydrolysis reactions described in Chapter 08).

A few metals, notably gold and silver, are so chemically inert that they rarely form compounds and occur as pure ('native') metal, usually as tiny flecks rather than in the fist-sized nuggets of popular myth. If the circulating fluid passes along fractures, precipitation may occur there so that the ores form within these veins rather than being dispersed throughout the neighbourhood. These fractures are usually filled mostly by undesirable minerals such as quartz, forming a vein that is mostly useless material but will be economic to mine for gold provided it contains more than about ten parts per million.

The origin of hydrothermal ore deposits associated with igneous intrusions is summarized in Figure 11.6. Other hydrothermal ores occur at ocean spreading axes at vents such as black smokers (Chapter 05). At present it is far too expensive to mine these on the sea floor, but they have been mined in ophiolites, notably on Cyprus, which was the main source of copper for the classical world.

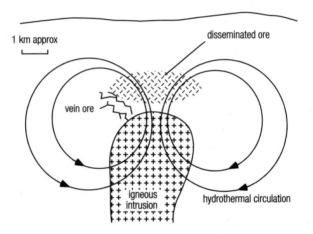

figure 11.6 Cross-section showing the precipitation of ore bodies from circulating hydrothermal fluids. The convection system is driven by the heat of the igneous intrusion.

Some hydrothermal ore deposits are nothing to do with igneous intrusions at all. For example, the lead and zinc ore bodies in the Mississippi River basin, in Ireland and in the Pennines of England occur within limestones, and appear to have precipitated from warm mineralized fluids expelled from deeply-buried sandstones or shales. Warm groundwaters are also suggested to explain the concentration of uranium into ore bodies in the Athabasca Basin of northern Saskatchewan, Canada and in the Northern Territory of Australia.

Ore exploration

Exploration for ore bodies can be done in many ways. If the ore body extends to the surface, then it may be discovered by an old-style prospector. Even if the economic part of an ore body is not exposed, hydrothermal circulation may have recognizably bleached or otherwise discoloured the overlying rocks at the surface. Surface effects may also be discovered by analysis of visible and infrared images recorded by satellites or airborne surveys. Other exploration techniques include mapping the magnetic field, often by an airborne survey, and looking for anomalous magnetic features that may be caused by metallic ores. When a potential ore body has been located, it must be drilled to assess the grade of the ore and confirm its extent. Only about one in a thousand promising discoveries has all the attributes required to be turned into a successful mine.

Mining of ores

Deep mining is the standard method for most ore bodies (Figure 11.7) but hydrothermal copper deposits tend to be so extensive that they are usually exploited by surface mining, also known as quarrying or 'opencast' working (Plate 12). The deeper the hole grows, the more expensive this technique becomes, because the sides of the quarry must be expanded to prevent the walls from collapsing. The only effective means to ensure this is to cut the walls back into a series of steps. Thus a progressively greater ratio of unwanted 'overburden' must be removed every time the floor of the quarry is lowered (Figure 11.8).

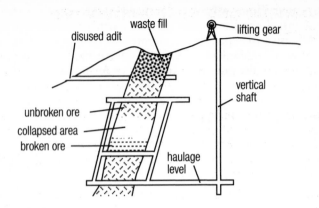

figure 11.7 Cross-section through a deep mine, exploiting an inclined ore body. An adit is a horizontal or gently inclined tunnel, used in this example to gain access to the upper part of the ore body, which has now been exhausted. Ore is now being extracted through deeper tunnels and raised mechanically up a vertical shaft.

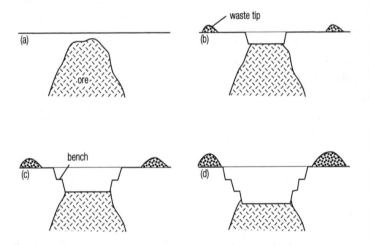

figure 11.8 Progressive stages in 'surface mining' or quarrying of an ore body seen in cross-section, showing how the sides of the quarry are cut into a series of steps or 'benches' for stability. Each time the quarry is deepened, a greater amount of overburden must be removed. Eventually it becomes uneconomic to continue. In this example, before the quarry can be deepened beyond stage (d) the waste tips would have to be moved.

Diagenetic and sedimentary ore bodies

Not all ores are formed as a result of hot fluids. Some form during low-temperature diagenesis of sedimentary rock, if the chemical conditions in the water permeating it are correct. A good example is provided by the ironstones of Northamptonshire and Cleveland in England. These are shallow marine sandstones of Jurassic age that are cemented by iron carbonate (siderite $FeCO_3$), sufficiently easy to access and of sufficient grade to have been an important ore until the latter part of the twentieth century.

Today, most of the global production of iron comes from another type of sedimentary deposit, consisting of alternating layers of silica and iron oxides and known as **banded iron formations** or **BIFs**. Virtually all BIFs are between 3.8 and 1.8 billion years old, and are believed to have formed on the floors of shallow seas when (as you saw in Chapter 02) there was much less free oxygen around. BIF ores have been mined in the Lake Superior district of the United States since the nineteenth century. About 15 per cent of the global production of BIF iron is from Australia, notably the Hamersley Basin in the north of Western Australia where it is mined in huge open pits.

Other ores are an end product of chemical weathering. The outstanding example of this is bauxite, the only globally significant ore of aluminium. This is produced in humid tropical conditions when hydrolysis gets to work on clay minerals. For example, kaolinite, produced by the chemical reaction shown in Chapter 08, breaks down to bauxite as follows:

$$Al_2Si_2O_5(OH)_4 \quad + \quad 5H_2O \ = \ 2Al(OH)_3 \ + \ 2H_4SiO_4$$

| kaolinite (a clay mineral) | water | bauxite ore | a form of dissolved silica |

Australia, Guinea, Brazil, Jamaica, China and India are the world's six major producers of bauxite ore. The only other important ores produced in this way are of nickel, found as insoluble silicate residue in soils formed by humid tropical weathering of olivine-rich igneous rocks. About a third of the global production of nickel comes from such sources (mostly from New Caledonia, Indonesia, Cuba, Brazil and the Dominican Republic), but the majority is from sulphide ores associated with ultrabasic lavas and the Sudbury asteroid impact.

The remaining class of ores is **placer deposits**. These occur when heavy, metal-rich grains become concentrated by sedimentary processes. Gold is the most obvious example. All too often, in 'them there hills' where the gold is supposed to be, it occurs as tiny flecks in veins far too diluted by barren quartz to be worth mining. However, streams draining this area, and therefore transporting locally derived sedimentary grains, can act to concentrate the gold. This is because grains of gold are considerably denser than normal sand grains and will tend to lag behind in the bed-load. When panning for gold, a prospector tries to concentrate the gold still further, by swirling a few handfuls of streambed sand or gravel in a basin, hoping to wash away the sand and leave the gold behind.

Most commercial gold production comes not from present-day streams, but by mining ancient placer deposits. The world's most prolific gold-producing area is the Witwatersrand Basin in South Africa, where gold is found in pebbly and gravelly sediments that appear to have been deposited by braided rivers about 2.5 billion years ago. The Witwatersrand conglomerates are notable also for containing placer grains of uraninite, an oxide of uranium (UO_2). Like BIFs, placer uranium ores are an indicator of the low level of atmospheric oxygen in the distant past. Placer uranium ores cannot form today, because under present conditions uranium forms the compound UO_3, which goes into solution rather than forming solid grains.

Non-metallic raw materials

Some non-metallic raw materials are sufficiently valuable to be mined underground like ores. These include fluorspar (CaF_2), which is needed for aluminium smelting and as a flux in steel making, and various evaporite minerals such as gypsum ($CaSO_4 . 2H_2O$), which is used in plaster and as a filler for textiles, and potash, a general name for potassium salts (e.g. KCl and $K_2SO_4 . 2MgSO_4$), which is used in fertilizers and other chemicals.

Modern industrial society requires other materials in bulk too. The most obvious are: clay and shale for bricks; limestone for cement, concrete and roadstone; sand and gravel for concrete, roadstone and various construction purposes; and crushed hard

figure 11.9 An opencast clay pit in County Durham, England. Deltaic mudrocks and silts are extracted and mixed in the right proportions to make bricks. Coal is a useful subsidiary product from this site. The digger is standing on the top of a coal seam, and another seam can be seen about 4 m up in the far wall.

rock in general for roadstone. These are extracted by quarrying. Sometimes a happy occurrence of mudrocks and coal seams gives two economic products from a single operation (Figure 11.9).

Making good the damage

In Britain, sand and gravel come most readily from shallow quarries on the flood plains of lowland rivers. Large quantities of these materials were deposited at the end of the last glaciation, when these rivers were transporting a much greater bed-load than at present. After use, these gravel pits usually become flooded, and may be used for watersports, or turned into nature reserves.

Deeper disused quarries and mines pose more problems. Many quarries are used as 'landfill' sites, for disposing of domestic and industrial refuse. With sensitive landscaping an infilled quarry may blend perfectly into the surrounding countryside, though it annoys geologists if no rockfaces are preserved. Controls are required to limit what types of waste may be buried in specific localities. For example, if the quarry is in a permeable limestone, then nothing should be buried there that might contaminate the groundwater. Mines are less of an eyesore, especially if the heaps of mine waste or 'spoil' are sculpted to blend into landscape. Old mines are rarely backfilled and the most lasting legacy of mining on an area may be subsidence. Above shallow mine workings, buildings and roads on the surface may become damaged centuries after a mine has closed.

Water resources

One other great natural resource that is taken from the ground in many places is water. Water may fill the pore-spaces of permeable rocks, and can be pumped from wells drilled into these natural underground reservoirs. If the permeable layer is capped by an impermeable layer, the groundwater may be under sufficient pressure to rise to the surface without pumping, in a so-called artesian well (Figure 11.10). When the fountains in London's Trafalgar Square were first constructed the water gushed out under artesian pressure, though nowadays extraction of groundwater has lowered the water table so the fountains have to be fed by pumped water. Many of the oases in the Sahara desert are where artesian water in a sandstone layer reaches the surface.

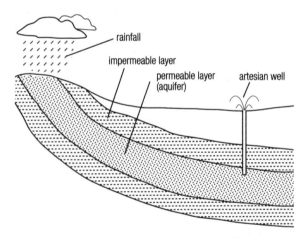

figure 11.10 Cross-section showing how a permeable layer (known as an aquifer) soaks up water where it is exposed at the surface. This water is prevented from escaping by impermeable rock above, but water will rise to the surface under pressure if a well penetrates into the aquifer.

Groundwater is essentially a renewable resource because the groundwater reservoirs are recharged by rainfall. However, in dry regions such as Arabia and the southern United States, groundwater is being extracted much faster than its recharge rate.

Perhaps the most priceless natural resource of all that the Earth has to offer is living organisms. Traces of life can be found in many ancient rocks. We have already discussed how coal and hydrocarbons are made from organic remains, and how limestone is usually made from the shells of living organisms. In the next chapter we will look at some of the more interesting or important organisms that can be found fossilized in rocks, and at the history of life on Earth.

12

past life and fossils

In this chapter you will learn:
- about the history of life on Earth, including mass extinction events
- how the remains of dead plants and animals can become preserved as fossils
- about the kinds of information fossils can give to the geologist.

For many people, **fossils** and geology are inseparable. There is no denying the thrill of discovering the remains of an organism that lived hundreds of millions of years ago. The study of fossils has taught us most of what we know about the history of life, and has provided much of the evidence in support of the theory of **evolution**. Fossils are also important because signs of life in sedimentary deposits, whether the remains of the organisms themselves or traces such as burrows and the like, can be just as valuable as bedforms, grain size, and so on, in helping us determine the nature of a past environment. Moreover, whereas most characteristics of deposits in any particular environment are the same irrespective of age, identification of varieties of fossil organisms provides the single most useful way of dating a rock sequence. This is because many species have survived for only limited periods, of the order of a few million years in the most useful cases.

The origin of life

The presence of life on Earth used to be a great mystery. However, knowing what we know now it would be a surprise to encounter an Earth-like planet where life had *not* developed. The chemical elements required for life can be found in abundance in the oceans, and even more abundantly in hot fluids escaping from hydrothermal vents such as black smokers (Chapter 05). Experiments in the 1950s showed that amino acids (from which proteins can be built) can be made by passing electric sparks through a mixture of ammonia, methane and water. Lightning discharges into the Earth's early oceans could therefore have begun the sequence of chemical processes that ended up with life. Amino acids have also been found in carbonaceous meteorites, so an alternative possibility is that these building blocks of life were delivered by meteoritic (rocky) or cometary (icy) debris falling onto the still-cooling and sterile surface of the young Earth.

There are many steps on the road from simple amino acids to complex molecules capable of self-replication (such as DNA) and thence to cellular life. However, geological time is so long and the number of molecules in the oceans so vast that there is no need to call on divine intervention to account for how life began.

The first microbes

It seems most likely that life on Earth began maybe as long as 4 billion years ago in the form of microscopic single-celled organisms (microbes), like bacteria or a less familiar group called archaea, beside underwater hydrothermal vents where they would have been nourished by the abundant supply of chemical energy. Chemical reactions occurring across the skin of tiny bubbles could have led to the origin of the first cell membranes. There are varieties of microbes still around today obtaining their energy chemically and able to survive only in oxygen-free conditions, which are probably little changed from the primitive cells that developed beside those ancient hot vents. Today they are most noteworthy for their function in breaking down raw sewage; a rather unprestigious role for the closest relatives of Earth's first living organisms.

Once bacteria-like organisms had arisen, new varieties evolved and migrated to less chemically rich regions where energy had to be extracted from sunlight (by photosynthesis). The simplest photosynthetic organisms today are the cyanobacteria, formerly called blue-green algae. As you saw in Chapter 02, once these organisms began to spread and multiply they also began slowly to change the chemical balance of the atmosphere, in particular liberating oxygen, which was at only one per cent of its present level about 2 billion years ago.

The earliest visible traces of life are cells and filaments of cells, found as fossils in chemically precipitated chert and carbonate deposits known from radiometric dating of associated volcanic rocks to be 3.5 billion years old. These fossils are only about a hundredth of a millimetre across, and so can be discovered only by painstaking microscope surveys on thin, translucent, slices of rock. Some scientists contend that these structures were produced chemically rather than biologically, but all agree that if they do represent real cells then they were very simple, like cyanobacteria, lacking a nucleus and other internal structures.

However, less contentious fossil structures large enough to see with the unaided eye date back almost as far. These are metre-scale mat-like or dome-like laminated features within calcareous sediments in the Pilbara region of Western Australia, thought to be between 3.4 and 3.5 billion years old. They resemble modern-day stromatolites, sometimes known as algal mats, that are held together by strands of cyanobacteria. Fossil stromatolites became abundant about 2.3 billion years ago, and remained so for nearly 2 billion years until their decline, presumably as a result of grazing by animals and competition with more advanced plants.

More complex cells

The next step up the evolutionary ladder from microbes and algal mats came with the appearance of more advanced cells of the kind known to biologists as eukaryotes, to which plants, animals and fungi belong. In contrast to archaea (from which they are probably descended) and bacteria, eukaryotes have thick cell walls, a nucleus, chromosomes, and various other internal structures. The earliest known eukaryote fossil cells are about 2 billion years old, and are probably single-celled planktonic plants that lived in the sunlit upper waters of the ocean.

Multicellular organisms

Fleshy multicellular marine algae appeared about 1.2 billion years ago, but the first multicellular animals do not appear in the fossil record until approximately 900 million years ago. These had soft, worm-like bodies and so are imperfectly preserved as fossils. It is not until 600 million years ago that a wide variety of forms emerged, in both shallow and deep-water environments. These were still soft-bodied organisms, but an important change happened 540 million years ago, when animals developed hard parts such as shells that do not decay upon death and which therefore fossilize easily. From then onwards the fossil record is rich, diverse and dominated by remains of animals with hard parts.

The interval of time since 540 million years ago has long been known to geologists as the Phanerozoic Eon. (If you are unfamiliar with the divisions of geological time, you may find it helpful to refer to Appendix 3.) Phanerozoic means 'visible life', and was so named because it is not until the beginning of the Phanerozoic, during the period known as the Cambrian, that sediments were deposited in which fossils are abundant and easily recognized.

The time prior to the Cambrian is often informally described as Precambrian. It is divided into three eons: the Hadean Eon from the Earth's origin until 3.9 billion years ago; then the Archean

Eon until 2.5 billion years ago; and finally the Proterozoic Eon extending from the end of the Archean to the start of the Phanerozoic. 'Proterozoic' means 'earlier life', though as we have seen, we can now trace the earliest life way back into the Archean.

The major animal phyla appear

Most of the major divisions (phyla) of the animal kingdom arose during or soon after the Cambrian. These include:

1 two phyla with shells made of calcium carbonate (Figure 12.1), namely molluscs (some with two shells and some with one shell) and the less-familiar brachiopods (two-shelled organisms, but with body-plans entirely different from molluscs);

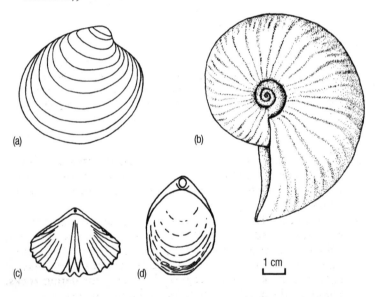

1 cm

figure 12.1 Bivalve (a) and ammonoid (b), two distinct classes of the mollusc phylum. Two brachiopods (c) and (d), which although at first sight similar to bivalve molluscs belong to a different phylum.

2 a phylum known as the arthropods, with jointed external skeletons (exoskeletons) made of a resilient organic material called chitin, most famously represented in the fossil record

by trilobites (Figure 12.2) whose present-day distant relatives include crustaceans (e.g. crabs and lobsters), spiders and insects;

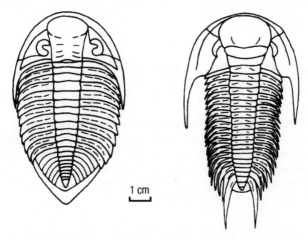

1 cm

figure 12.2 Two trilobites, representatives of a kind of arthropod that first appeared in the Cambrian and became extinct during the Permian. These two examples show the complete articulated exoskeleton, except that the fragile antennae and legs are missing. Compound (insect-like) eyes can be seen on either side of their heads.

3 a phylum known as the echinoderms, characterized by a multi-element skeleton of calcium carbonate, often with five-fold radial symmetry, that includes sea urchins, star fish and crinoids;

4 the hemichordate phylum, obscure and unimportant today, but represented in the Paleozoic by a group of floating colonial organisms known as graptolites (Figure 12.3), whose rapid evolution and widespread distribution in marine sediments make them exceptionally useful for dating rocks, especially during the Ordovician and Silurian periods.

Another important phylum, the cnidarians (or coelenterata), which includes the jelly fish, appeared in the late Precambrian. These are very rare in the fossil record until about 460 million years ago when corals first appeared. The reason that corals are well represented in the fossil record is that the soft coral polyp encloses itself in a hard calcium carbonate cup. This is readily preserved after death, especially in the case of massive colonial corals in which adjacent cups reinforce each other.

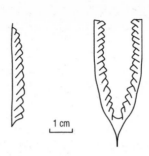

figure 12.3 Two graptolites. These two- and one-branched forms are characteristic of the graptolite species abundant in the Ordovician and Silurian, respectively. Only the hard parts are shown here, consisting of a series of cups arranged along one or more branches. Before fossilization these were made of a protein called collagen. In life, a separate animal lived in each cup, catching food particles as the colony drifted through the water.

The chordate phylum, which includes the vertebrates (and therefore humans) was around from the early Cambrian, although initially it is insignificant in the fossil record. However, a flourishing fauna of jawless fish developed about 500 million years ago, and bony fish with jaws followed about 80 million years later.

Vascular plants and life on land

About 420 million years ago 'vascular plants' with tubes to transport water and nutrients began to colonize the land, and were shortly followed by insects and other arthropods. Some fish developed the capability to breathe out of water, and about 360 million years ago some of their descendants evolved into the first primitive amphibians (living on land as adults but breeding in fresh water), followed within 20 or 30 million years by the earliest reptiles.

Animal life on land was dominated for a long time by reptiles, which, unlike amphibians, could breed on land as well as live there. The most famous are the dinosaurs that flourished throughout the Mesozoic era (encompassing the Triassic, Jurassic and Cretaceous periods). At this time, the swimming reptiles such as ichthyosaurs (Figure 12.4) and plesiosaurs were among the largest marine creatures. Other relatives of the dinosaurs, the pterosaurs, took to the air.

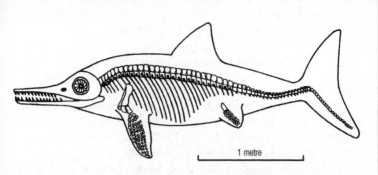

figure 12.4 An ichthyosaur, a marine reptile from the Mesozoic era. Usually only the bones and teeth survive, but in this example an impression of the skin is preserved, which enables details such as the dorsal fin and the upper part of the tail to be shown in the outline around this skeleton.

Mammals were present on land for most of the Mesozoic, though they were small and insignificant creatures, known mostly from their teeth, which are the only parts of such small-boned creatures to fossilize well. The earliest known bird, Archaeopteryx, is of late Jurassic age (about 170 million years old). This magpie-sized creature, known from fossils found in the fine-grained Solnhöffen Limestone of Germany, has teeth and dinosaur-like feet and would probably have been dismissed as a winged reptile but for the impressions of its plumage within the rock.

The first flowering plants appeared as recently as 100 million years ago, midway through the Cretaceous, and did not surpass the conifers in importance for a further 30 million years or so. Back in Carboniferous times (360–300 million years ago), the trees responsible for forming most of the world's coal deposits belonged to neither of these groups, instead vegetation was dominated by giant relatives of the club mosses (lycopods), which today are small and inconspicuous.

The mammals take over

It was not until the dinosaurs disappeared at the end of the Cretaceous Period (65 million years ago), at the so-called **K-T boundary**, that mammals had their chance to diversify, which led to them becoming the dominant group of land animals. They also spread into the sea, with the ancestors of modern whales first appearing about 50 million years ago.

The fossil record of human ancestors is not good. Complete skeletons are unknown, and most of the story has to be pieced together from skulls (usually incomplete) and other bone fragments. The first human-like apes ('hominids') appeared about 6–7 million years ago. The first bones recognized to be from species belonging to our own genus, Homo, are about 2.4 million years old and our species Homo sapiens does not appear until about a quarter of a million years ago.

This timescale should make it clear to you that movies and cartoons showing humans hunting dinosaurs (or vice versa) are very wide of the mark indeed, because they overlook the matter of a 60-million-year gap! On the other hand, that other well-known extinct animal, the mammoth (essentially a hairy elephant) was very much a human contemporary. Mammoths feature prominently in cave paintings, and the last ones died out as recently as 12 000 years ago. It is debatable to what extent their final demise was due to hunting or to the loss of their habitat because of climate change.

Mass extinctions and adaptive radiations

But why did the dinosaurs become extinct at the K-T boundary, after having been so successful for so long? This is a question that has taxed geologists and palaeontologists (those who study fossils) for many years. Various explanations have been proposed, including rapid climate change (perhaps caused by catastrophic methane escape from seafloor sediments, or by sulphur dioxide aerosols from flood basalt eruptions), and genetic damage by cosmic rays from a nearby supernova. Probably several factors were at play, and the dinosaurs appear to have been in decline for at least a few hundred thousand years before the probable knockout blow was delivered when a 10-km-diameter asteroid or comet collided with the Earth. The impact site of the main fragment has been identified as the 200-km-diameter Chicxulub crater on the Yucatan coast of Mexico. This vast structure is buried beneath more recent sediments and was discovered by geophysical techniques in the 1980s. A few other buried craters of corresponding age (as close as can be told within uncertainties of the measurements) have since been found, notably the 24-km-diameter Boltysh crater in the Ukraine and the smaller Silverpit crater beneath the North Sea about 100 km off the Yorkshire coast of England (Figure 12.5), suggesting that the incoming object broke into several fragments before it struck.

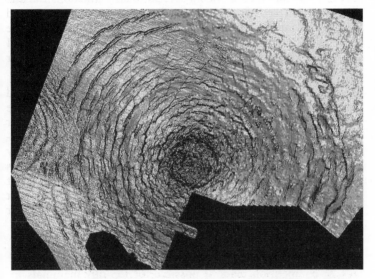

figure 12.5 20-km-wide view of the Silverpit structure, revealed by seismic surveying. This is buried below more than a kilometre of Cenozoic sediments, and may be an impact crater of the same age as the K-T boundary. The black areas are unsurveyed.

The main impact would have injected a vast amount of dust into the atmosphere. Supplemented by smoke from widespread burning, this would have reduced the sunlight reaching the surface to such an extent that photosynthesis became impossible. Plants would have died, and the whole food web depending on them would have fallen apart. Thanks to surviving seeds and spores, recovery of the vegetation on land and of marine planktonic plants would have begun within a few years, but many groups of animals and some plants became extinct. Whatever the cause, the effects were much wider than merely the demise of the dinosaurs. Many important marine groups vanished, notably the ammonoids (Figure 12.1b), reef-building molluscs called rudists, and the large marine reptiles. Most species of marine planktonic animals of the kind known as foraminiferans disappeared too, although the group as a whole recovered.

Because it was global in extent, and has effects that can be seen among organisms living in virtually all environments, what happened at the K-T boundary is described as a **mass extinction** event. The fossil record shows several mass extinctions during the Cambrian Period, and a more dramatic one at the close of the Ordovician Period. Because there was no life on land yet, these mass extinctions are purely marine. However, a mass extinction event at the end of the Permian Period (marking the end of the Palaeozoic Era) wiped out whole groups of land and marine animals and, biologically speaking, was an even more severe event than the K-T mass extinction.

The causes of these earlier mass extinctions are uncertain, and it is unlikely that all were caused by impacts. What is certain is that after each mass extinction those varieties or organisms that had survived were able to take advantage of the demise of their competitors to adapt, through evolutionary change, to new lifestyles. This is described as adaptive radiation.

The sudden blossoming of the mammals at the start of the Cenozoic Era is a good example. The disappearance of the dinosaurs at the K-T boundary meant that there were no longer any large land-dwelling herbivores or predators. The sudden lack of competition made it possible for the tiny mammals to diversify, and within a few thousand generations some groups that had moved into new environments were well on the way to evolving into different species better suited to take advantage of the new opportunities. A classic case of the meek inheriting the Earth, if ever there was one!

How fossils form

A fossil is any sign of a past organism entombed in rock. Preserved footprints and burrows count as fossils, and these are termed trace fossils. On the other hand, an impression of the body where it came to rest after death, or some part of the body that has actually been preserved, is described as a body fossil. It is rare for fleshy material to survive – it is usually broken down by bacterial decay or eaten by scavengers before there is time for it to be buried, which is why the vast majority of fossils are the bony, shelly, or other hard parts of an organism.

To follow the stages by which an organism can be turned into a fossil, we will take the example of a bivalve mollusc, like a cockle or some similar two-shelled shellfish probably familiar to you from visits to the seaside. To become a fossil is a long and involved process, and is an unlikely outcome for the majority of individuals.

The most obvious requirement for an organism to end up as a fossil is that it must become buried. There are many chances that can prevent this happening. If our mollusc has the misfortune to be killed by a predator, then the soft parts will be eaten. The predator will probably have to break at least one of the shells to gain access to the flesh, so it is unlikely that both shells will survive intact. Even if they do, the two shells may become separated. On the other hand, if the organism dies a natural death, although the soft parts of the body will almost certainly rot away neither shell is likely to be damaged. The shells are held together at the hinge by fibrous tissues, which are among the last to rot so there is a reasonable chance that the two shells will remain held together until they become buried.

A lot depends on where the creature is when it dies. If it is the sort of bivalve that lives in rivers, then after death the empty shells are likely to be rolled downstream and will pretty soon be broken up. If it is a marine bivalve that lived and died on a rocky shore it will probably be smashed in the next major storm. If it dies on a beach, it would get rolled around by the waves, the two shells would probably separate and become broken and abraded, ending up as unrecognizable fragments of shell of the sort that make up some kinds of limestone. Only if the organism dies in a low-energy environment, somewhere quiet enough for muddy sediment to be accumulating, does it stand a reasonable chance of escaping mechanical breakdown of the sort described above long enough to become buried by sediment.

However, there is a very important exception to the previous argument, which is that many species of bivalve (including cockles) spend most of their time burrowing within the sediment. If one of these dies in its burrow, then it is already buried, and the two shells will be held together, confined by the surrounding sediment, even after all the connecting tissues have rotted away. An instant fossil, you might think, but this is far from the end of the process (Figure 12.6). Very few fossils, and only relatively recent ones, occur as unchanged shell material.

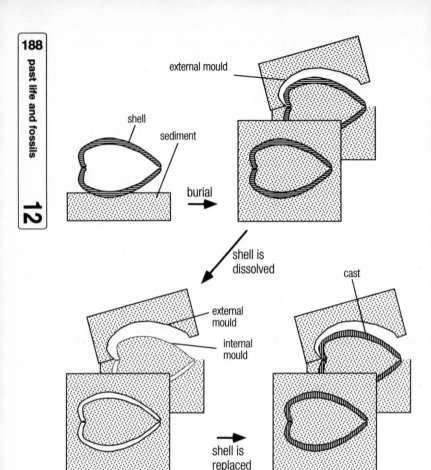

figure 12.6 Stages in the fossilization of the hard parts of an organism. A bivalve mollusc is shown here (in cross-section), but the same processes can affect any kind of shell, tooth or bone.

The first thing to happen to a buried bivalve, minus all its soft parts, is usually that some sediment will find its way inside the shells. Once the sediment has hardened (by the diagenetic processes described in Chapter 09), the rock contains an impression of the outside and the inside surfaces of the shells, known as an external mould and an internal mould respectively. Even if, as often happens, the shells are subsequently dissolved away, these moulds remain, and many fossils are of this type (Plate 13). If the shell survives, usually it recrystallizes in some way, so that its original microscopic crystalline structure is lost. It often happens that the whole shell is dissolved by ground water, but if conditions revert, the shell-shaped void in the rock can be re-filled by precipitation, to produce a cast whose mimicry of the surface texture of the vanished shell is limited only by the quality of the impression that it left in the sediment. It can be very difficult to distinguish a cast from a recrystallized shell if these are of the same composition (calcium carbonate, in the case of molluscs).

The steps described above apply to most types of fossil, with various changes of emphasis. An ammonoid has only a single shell, so the complete hard parts stand more chance of being preserved intact than in the case of a bivalve mollusc, but on the other hand ammonoids did not live in burrows, so in that respect the chances of becoming fossilized are reduced. Sea urchins and other echinoderms have a skeleton consisting of a large number of fragile plates, which are easily separated after death, so low-energy conditions and rapid burial are more important than for bivalve molluscs (fortunately many kinds of sea urchin are burrowers, so these can be found beautifully preserved). Trilobites and other arthropods have segmented skeletons, so rapid burial is especially important for these, too, if the hard parts are to remain intact; moreover, the chitin of which the skeleton is made does not survive over geological time, and must be replaced by some form of mineral if anything other than a mould is to be preserved. In vertebrates, the teeth (made of a dense crystalline form of calcium phosphate) are the most robust elements, so these are the bits that are most commonly found as fossils. Bones (in life, a porous growth of calcium phosphate) are less commonly preserved, and when they are, their porous structure has usually been invaded and replaced by new minerals precipitated from groundwater.

Exceptional preservation

The foregoing does not mean that flesh cannot be preserved, though it is rare. Examples include complete 20-thousand-year-old mammoths frozen in ice in Siberia, and 100-million-year-old fish in northern Brazil, which were killed by highly saline waters that caused phosphate minerals to be precipitated inside their bodies within a few hours of death, so that the fine scale cell structure is sufficiently well-preserved to warrant study using high-powered electron microscopes.

Perhaps the best-known circumstance for a whole organism to be preserved is when it has the misfortune to become trapped in tree resin, which then hardens to form amber. Delicate insects and even small frogs can be found within amber. It is even true that the blood cells from a blood-sucking insect's last meal can end up being preserved in this way, though cloning a dinosaur from the DNA within these cells, as in the film *Jurassic Park*, remains a very distant prospect.

Unusual preservation may also occur when environmental conditions change so suddenly that a whole living community is wiped out. A classic example of this is found in the middle-Cambrian Burgess Shale of British Columbia, Canada, where animals living on a muddy sea floor were now and then transported by submarine landslides into deeper water, lacking in oxygen, where they died and were immediately buried. The lack of oxygen kept scavengers at bay and inhibited decay, with the result that entirely soft-bodied animals are preserved, as flattened organic residues. This sort of 'mass mortality' deposit is important because it gives us an unbiased snapshot of a living community. In the case of the Burgess Shale, only about one fossil in 20 is a hard-shelled organism, which is a salutary reminder of how heavily most of the fossil record is skewed in favour of robust organisms with easily preserved hard parts.

Plant fossils

Plants have no shells or bones, so you might expect them to make rather poor fossils. However, land plants have more rigid cell walls than animals, made of cellulose. As a result, woody tissues and even leaves tend to retain their form after death and burial, long enough for the spaces inside the cells to be filled by minerals, such as chert, that precipitate from groundwater. This is how petrified wood is formed.

An alternative mode of preservation is when plant material is buried in waterlogged soil or mud with a low oxygen content. In this case the leafy or woody material decays to carbon or a carbon-rich compound.

A major problem in understanding fossil plants is that it is very uncommon to find a whole plant preserved. Roots (as you might expect) are relatively common, but these are rarely found still in association with leaves or cones. This has led to different parts of the same plant being given different fossil names!

On the microscopic scale, pollen and spores are often well preserved, because their coats are made of sporopollenin, a substance that is remarkably resistant to decay. In sediments up to a few million years old, when many plant species were the same as those today, or their close relatives, analysis of pollen is a very important way of establishing the environmental conditions. For example, changing tree communities in northern Europe towards the end of the last glaciation are well documented by the pollen record, which indicates climatic fluctuations between warm and cold superimposed on the general warming trend.

The use of fossils in stratigraphy

It is unusual for sediments to contain minerals that have grown within them whose age can be dated by the radiometric dating techniques we met towards the end of Chapter 02. If a sediment is interbedded with a series of lava flows or layers of volcanic ash, then radiometric dating of these volcanic rocks can be used to bracket the age of the sediment in between. However, convenient layers of igneous rock are often absent from sedimentary sequences, and anyway radiometric dating requires sophisticated laboratory facilities that a geologist is unable to call on while in the field. This is where being able to identify fossils pays off. If you already know the time span over which a particular kind of fossil organism was alive (because somebody else has been able to determine this in a location where fossil-bearing sediments *are* interbedded with radiometrically datable volcanic rocks), then you can tell the age of a rock merely by recognizing the fossils within it.

To take a simple example, trilobites are pretty distinctive (Figure 12.2), and there is not much risk of mistaking a trilobite for something else. You have already seen that trilobites first appeared during the Cambrian and became extinct during the Permian. This means that if you find a trilobite in a rock then its age must lie within these limits. This is a period of about 300 million years, which is of course rather a long time. However, during this period many different species of trilobite evolved and then died out, with an average species lifetime of around 10 million years. The two trilobites shown in Figure 12.2 are clearly different from each other and although they are both trilobites, these two are not very closely related. The differences between closely related species are usually more subtle, and it requires well-preserved specimens and an expert's eye to make an unequivocal diagnosis enabling the age of the sediment to be pinned down to within 10 million years.

It is possible to improve on this if, in the same layer of sediment, you are fortunate to find fossils of two species that have overlapping age ranges. This can be seen with reference to Figure 12.7. If you found only species C in a bed of rock it would not pin the age down very closely, because this species was alive from 500 to 467 million years ago. Species B on its own would be more diagnostic, because it was around for a shorter period (493–478 million years ago), and species D on its own would give an even more precise age (470–460 million years). However, if you found species B in the same bed as species A, then the age of this bed could be pinned down to the 2-million-year period (480–478 million years ago) where the age ranges of these two species overlap. An almost equally precise date could be inferred if species C and D were found in the same bed.

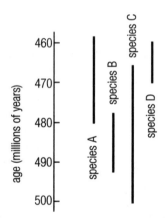

figure 12.7 The age ranges of four species of fossils (heavy lines). See text for discussion.

The above shows how we can infer quite precise ages of sediments, especially if we can use several fossils in combination. These principles can be extended to match up the ages of sediments deposited in areas that are remote from each other. For example, if species A and B were found together in North America and in China we would know that these two unrelated deposits were of almost exactly the same age (this would be true even if we did not know the absolute ages from radiometric studies). Correlating between rock sequences in this way is called **biostratigraphy**.

Some fossils are better than others for biostratigraphic correlation. The best ones are called zone fossils, or index fossils. A good zone fossil:

- has easily recognizable features that enable it to be distinguished from other species
- underwent rapid and irreversible evolutionary change
- is numerically abundant
- lived in a wide range of environments
- lived across a wide geographical area.

Trilobites satisfy the first three points, but usually fail on the last two. Most trilobite species had fairly specialized lifestyles, and not being good swimmers their geographic ranges were limited (for example, entirely different species occurred at the same time on opposite sides of major oceans). The best zone fossils tend to be organisms that swam or floated passively in the oceans and whose hard parts, upon death, sank to the bottom and can therefore be found in both deep- and shallow-water marine sediments. Prime examples are graptolites (especially useful in the Ordovician and Silurian), ammonoids (especially in the Jurassic and Cretaceous), and microscopic planktonic foraminiferans (notably during the Cenozoic).

There is one major pitfall to beware of when using fossils for dating, which is that fossils can be eroded from an old sediment, transported somewhere else and then reburied. 'Derived fossils' such as these will suggest erroneously old ages. However, the derived nature of a fossil can usually be recognized by signs of abrasion and other damage during transport. One way to be sure that a fossil belongs in the rock where it is found is if it is in 'life position', as is the case, for example, with a bivalve mollusc that died in its burrow or a coral that is still attached to the rest of the reef.

Now we have completed our overview of life on Earth, and seen how fossils can be used to correlate ages of rocks between different parts of the globe, we can turn our attention, in the next chapter, to global Earth history.

13 a short history of the Earth

In this chapter you will learn:
- about the Earth's evolution from a young semi-molten ball to a globe with a familiar pattern of drifting continents
- how the global climate has changed on a variety of time scales, and how the pace of change is being forced by human activities.

At the end of Chapter 02 you followed the story of the Earth's origin as far as its birth as a hot, young planet. You also saw how the atmosphere has changed over time. Now we will trace the other main events in the history of the Earth, and risk a few predictions about the future.

We will resume our story by imagining the young, hot Earth about 4.5 billion years ago. Most of the outer part is likely to be molten, because of the heat liberated by major impact events. The surface skin has cooled and solidified, but it is punctured every so often by upheavals when primordial gases escape from the interior, and by impacting debris from above. The magma ocean (as it is called) below the skin is convecting vigorously, and from time to time the convection currents rip the solid skin asunder, and a vast tract of cold, and therefore dense, skin founders beneath a fresh cover of magma. A terrible place, but eventually it will turn into our home.

The Earth's first crust

The solid skin of the infant Earth is not what most geologists would recognize as a planetary crust, because it is was merely the chilled equivalent of the magma beneath it, and had the same composition. However, during the next 100 million years or so, the magma ocean below the skin would have cooled down sufficiently to allow crystals to grow within it. The first would probably be a calcium-rich variety of feldspar known as anorthite. These would be less dense than the remaining magma, and so would tend to rise upwards, ultimately displacing the chilled skin to form the Earth's first true crust.

The remaining volume of the underlying magma ocean would now have a different composition, because of the loss of those elements that were concentrated into the anorthite crystals. We can now refer to this as the mantle. Deeper still, the process of core formation (see Chapter 02) was probably essentially complete by now.

None of the anorthite-rich 'primary crust' survives on the Earth today, but its equivalent can still be seen in the pale highland regions of the Moon. These are formed of a rock type called anorthosite, which is dominated by anorthite crystals.

Throughout this crust-forming process, the surface would have suffered from a continuing rain of impacts by Solar System debris. Although there were no planetary embryos left, it is clear from looking at craters on the lunar highlands that there were still plenty of small- to medium-sized planetisimals, capable of forming craters up to hundreds of kilometres across. Most would have been rocky (like asteroids) but some would have been icy comets, and opinion is divided as to what proportion of the Earth's water was delivered by cometary impacts at this late stage. There were even a few larger plantesimals whose impacts created 1000-km-wide basins on the Moon as recently as 3.9 billion years ago. The size of these largest impactors can be deduced by bearing in mind that collisions within the Solar System usually take place at speeds of a few tens of kilometres per second (not per hour!), and that a crater produced by such an impact is about 30 times the diameter of the impacting body.

Dating of lunar craters also shows us that the rate of impact events declined sharply after this time, and had reached something like its present level by about 3.8 billion years ago. This marks the near-exhaustion of the widely dispersed debris left over from the birth of the planets. Subsequent impacts have been mostly by stray comets or by rocky or iron-rich chunks scattered out of the asteroid belt by chance events. We know some of these hit the Earth as well as the Moon, because there are nearly 200 confirmed impact craters on the Earth. However, the vast majority of the Earth's impact craters have been buried by sedimentation, removed by erosion, or destroyed by tectonic activity.

To return to the Earth well before 4 billion years ago, with its newly formed primary crust of anorthosite and still in the throes of the early bombardment, it is likely that the mantle was convecting vigorously, even though by now it was mostly solidified. Above sites where convection cells diverged, the primary crust would be rifted apart, and partial melting of the upwelling mantle would yield basaltic crust similar to today's oceanic crust. As you saw in Chapter 04, oceanic crust is recycled at subduction zones. No Archean oceanic crust has survived, except for some debatable and deformed fragments caught up in collision zones. However, its presence can be inferred because partial melting of oceanic crust at subduction zones appears to have led to the growth of continental crust. Andesitic and other magmas above sites of subduction impregnated and overprinted the primary anorthosite crust, to the extent that no recognizable remnants of that have survived (Figure 13.1).

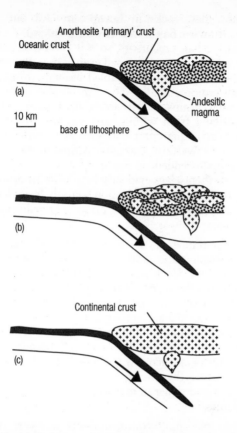

figure 13.1 Hypothetical cross-sections through the lithosphere about 4.2 billion years ago, showing how anorthosite primary crust could have been replaced by continental crust. In (a), andesitic magmas produced by partial melting of subducting oceanic crust are beginning to intrude the primary crust, and by stage (c) the process is complete.

The oldest known continental crust consists of metamorphic rocks about 3.8 billion years old, with the most famous examples being near Isua in south-west Greenland. The oldest unit here is the Amitsoq gneiss (apparently a metamorphosed granite), which is overlain by slightly younger metamorphosed sediments (containing tiny flecks of carbon that appear to be biological in origin) and volcanic rocks. Examples of similar age occur in north-west Canada (the Acasta gneiss), in South Africa (the Sand River gneiss) and Western Australia (the Narryer gneiss). The oldest rocks in Europe are 3.8 billion-year-old gneisses at Gruinard Bay in north-west Scotland. Metamorphic rocks must have had precursors, so it is certain that the earliest continental crust must have formed even earlier. Recent advances in analytical techniques have enabled individual mineral grains to be dated radiometrically. The age record is held by 4.4-million-year-old grains of zircon (a very hard and resistant mineral) collected from within early Archean metamorphosed sediments (in fact a quartzite) from Jack Hills within the Narryer gneiss terrane in Western Australia. Although the sediments in which these grains ended up are 'only' 3.6 to 3.8 billion years old, the grains themselves evidently began life by crystallizing in an igneous rock at least 600 million years before. Whether this rock was true continental crust or primary anorthosite crust will never be known.

With the appearance of the first rocks, we need place less reliance on speculation and analogy with other planetary bodies. Subsequent developments in Earth's history can be read directly, although the record is patchy at first. During the early Archean, the picture is one of relatively small 'continental' masses, only about 50 km across, being carried around by the rapid opening and closing of intervening areas of oceanic crust. This oceanic crust differed from today's because the mantle from which it was extracted was about 200°C hotter than now. As a result, rather then being basaltic in composition, it was poorer in silica and richer in magnesium, and was of a rock type known as komatiite.

The edges of the small continental masses were marked by volcanism, presumably because of subduction-related melt generation where they overrode the oceanic crust. Erosion in the interiors shed sediment into the oceans, and each time two continental masses collided the sediments and volcanic rocks caught up between were deformed, and deeper parts of the crust were metamorphosed.

Archean plate tectonics?

It would not be wise to regard the Archean opening and closing of oceans and the associated movement of continental masses as plate tectonics in the sense we know it today. Both continental and oceanic crust, and indeed the lithosphere as a whole, were probably thinner than now, and we do not know whether these components behaved like rigid plates as they do today.

The growth of cratons

Throughout the Archean Eon the volume of continental crust grew by igneous activity wherever oceanic crust was consumed. Between about 3.0 and 2.5 billion years ago, many of the hithero individual tracts of continental crust became welded together by collisions. Many of these amalgamations of crustal elements developed into regions that have remained stable ever since. These are described as **cratons** or **shields**. Each craton has two elements: the original tracts of continental rocks, which are loosely described as 'granite', and the deformed belts of volcanic and sedimentary rocks, collectively termed 'greenstone belts', that were caught up in the collision zone between each 'granite' block. Such an assemblage of crustal material is described as a granite-greenstone terrain.

The largest granite-greenstone craton is the Canadian or Laurentian Shield, which makes up most of the eastern half of Canada from the Arctic Ocean to the Great Lakes (extending south of the border into the Minnesota–Michigan region). Most of the rest of North America is younger rocks that became accreted around the shield as a result of subsequent deposition of sediments, volcanic activity and plate collisions. Each of today's major continents has a cratonic element within it.

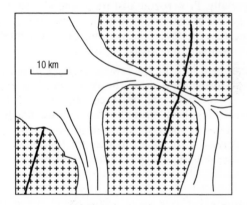

figure 13.2 Sketch map of part of a granite-greenstone terrain in Western Australia, near Marble Bar. This is part of the Pilbara Block, a 500-km-wide craton extending inland from Dampier and Port Hedland. 'Granites' are indicated by the cross ornament, and the trend of deformation structures in the intervening greenstone belts is shown by fine lines. Two dykes of Proterozoic age, post-dating the assembly of this craton, are shown by thick lines.

Australia has the Pilbara craton (Figure 13.2) and the Yilgarn craton to its south, Africa has the Kaapvaal and Zimbabwe cratons in the south and several others further north and west, Europe has the Baltic shield (Finland and adjacent parts of Russia), Asia's largest is the Aldan shield in the far south-east of Russia, and South American cratons occur in Brazil and Venezuela.

The Proterozoic Eon

Our story has now reached approximately 2.5 billion years ago, which marks the start of the Proterozoic Eon. Various authorities estimate that by this stage the volume of the continental crust had reached somewhere between 40 and 80 per cent of its present value. However, Archean cratons do not occupy anything like this proportion of today's continental surface. This is because the cratons represent only those regions of Archean crust that have remained stable ever since, and which happen not to be buried by younger sediments.

We will not attempt to survey Proterozoic history in detail. It is complex and poorly understood, but becomes easier to unravel as we approach the Phanerozoic (beginning 540 million years ago). Essentially, Proterozoic history is one of rifting apart and drifting together of cratonic blocks into new configurations, and gradual increase in the total volume of continental crust by means of volcanic and intrusive activity.

So far as can be told from the few deformed slivers that remain, the oceanic crust had now become basaltic rather than komatiite in composition, as a consequence of the mantle having become less hot. Some rifts never quite managed to split apart, and became sites of deposition for large volumes of sediment. Most of these sediments are types familiar today, except for the lack (or scarcity) of fossils. However, because the atmosphere was still poor in oxygen some sediments were kinds that do not form today. Notable among these are the banded iron formations (BIFs) that we referred to in Chapter 09. In fact, the commercially important Hamersley basin overlies the Pilbara craton south of the area shown in Figure 13.2.

Phanerozoic plate configurations

We will now wind the clock on to Cambrian times, beginning 540 million years ago at the start of the Phanerozoic Eon. Experts agree that well over 90 per cent of the Earth's continental crust had been created by this time. From now on, the record of the rocks offers enough clues for us to fit together a fairly complete story of events, although there is not space to trace it fully here.

It is clear that in Cambrian times the distribution of continental crust was radically different from today. Evidence for this includes study of palaeomagnetism, which is magnetism trapped within rocks (especially volcanic rocks) at the time of their formation. When rock cools down, it picks up an internal magnetization lying parallel to the Earth's magnetic field at the time. This enables us to determine both its orientation relative to the pole and its latitude at the time of its formation. It does not tell us whether it was in the northern or southern hemisphere, because, as we saw in Chapter 04, the polarity of the Earth's magnetic field has flipped many times in the past. However, this ambiguity can usually be resolved by back-tracking from the present location, provided that the same region includes suitably magnetized rocks from several intervals of geological time.

Other evidence comes from the similarity of rock types in continents that are now widely separated. Sediments are particularly useful in this respect, but intrusive rocks and deformed belts can also be informative.

Palaeomagnetism may tell us that two continents were at the same latitude at a certain time, but if they do not share any geological features of this age we may conclude that they were not joined. Palaeomagnetism does not tell us how far apart the two continents were, because their east–west separation is unconstrained. However, in the Phanerozoic there are abundant fossils, so if our two continents have similar fossils, then we can conclude that they were fairly close together, whereas if the fossils in the two regions are unrelated species it is likely that they were separated by a wide expanse of ocean.

To illustrate Phanerozoic Earth history, we will concentrate on the changing relationships between North America and northern and western Europe. In Cambrian times, most of the present continent of North America had been assembled. The bits still missing were various exotic terranes forming the present western coast (see Chapter 04), most of the present east coast from Newfoundland to the Carolinas, and also Florida. On the other hand, the continent was augmented by the presence of Greenland (firmly attached to what is now northern Canada), and by what is now Scotland and the north-western half of Ireland, which were somewhere close to Greenland. Clearly this continent was not identical to North America, and so it is generally given a different name: Laurentia, referring to the Laurentian shield at its heart.

Palaeomagnetic studies show that Laurentia lay astride the Equator during the Cambrian, and that the part that would later become North America was rotated about 90° clockwise relative to its present orientation (Figure 13.3).

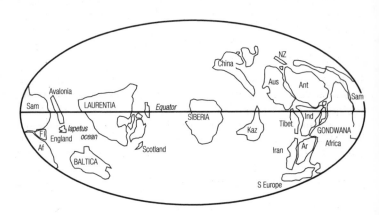

figure 13.3 Sketch map of the whole globe in the middle Cambrian, about 520 million years ago, showing the distributions of the continents. Present-day coastlines are drawn to aid identification, but do not correspond to shorelines at the time. Abbreviations: Af, Africa; Ant, Antarctica; Ar, Arabia; Aus, Australia; Fl, Florida; Ind, India; Kaz, Kazakhstania; NZ, New Zealand; SAm, South America. Many tracts of continental crust consisted of more than is implied by the names used here; for example, 'China' included much of south-east Asia, and 'Iran' included Turkey. The named regions within Gondwana as yet had no separate identities, and are outlined merely to aid recognition. England was part of Avalonia, which consisted of several microcontinents, most which are now exotic terrains on the east coast of North America. (Modified from various sources.)

The nearest other continent at this time consisted of what is today Scandinavia and most of Russia west of the Urals. This continent was built around the Baltic shield, and is referred to as Baltica. Palaeomagnetism shows that Baltica was roughly half way between the Equator and one of the poles, and it is most likely that it lay in the southern hemisphere.

To complete the Cambrian global picture, Africa, Florida, South America, Antarctica, Australia, India, Tibet, Arabia, 'Iran' and southern Europe were joined in a 'supercontinent' known as Gondwana, which straddled the Equator but appears to have been well out of the way of Laurentia. There were three smaller isolated continents consisting of what are now Siberia, the Kazakhstan region of central Asia, China and south-east Asia. Between Gondwana and Laurentia lay several microcontinental fragments, collectively known as Avalonia, which would later become the Carolinas, Newfoundland, Nova Scotia, the British Isles (except north-west Ireland and Scotland), and parts of northern France and Belgium.

Throughout the Cambrian, the geologies of Laurentia and Baltica have little in common. Fossils in marine sediments on the facing margins of the two continents are generally unrelated, whereas the unity of Scotland and North America is confirmed by the presence of identical species of trilobites and other animals. During the succeeding geological periods (the Ordovician and Silurian) the fossils in Laurentia and Baltica gradually become more similar. Clearly the ocean between the

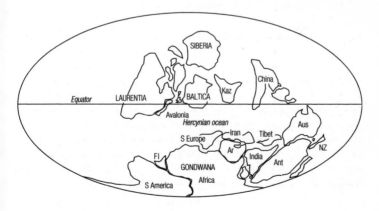

figure 13.4 Map in the style of Figure 13.3, showing the continents in the early Devonian, about 400 million years ago. See caption to Figure 13.3 for explanation. (Modified from various sources.)

two continents was becoming narrower, and the two continents had collided by the early Devonian (Figure 13.4). This now-vanished ocean is given the name Iapetus.

Various bits of Avalonia accreted onto Laurentia from the middle of the Ordovician (about 470 million years ago) onwards. English fossils became more similar to Scottish ones during the Ordovician and Silurian, showing that the English terrane was sliding into the Laurentia–Baltica collision zone. Its speed looks prodigious if you compare Figures 13.3 and 13.4, but it was only about 10 cm per year, which is comparable with present-day speeds of relative plate movements.

The geology of the Avalonian terranes contains many indicators of their histories. For example, many of them contain andesitic volcanic rocks of the kind generated above subduction zones, showing their essentially island arc character prior to accretion onto Laurentia. The Ordovician volcanic rocks of the English Lake District are a good example of this, and the Southern Uplands of Scotland consist of a deformed stack of sedimentary rocks that accumulated in a subduction zone trench. The edge of Laurentia that bore the brunt of the collisions was deformed by folding and thrusting, and a number of slices of the Iapetus ocean floor survive as ophiolites thrust onto the continent, notably several in Newfoundland.

By the end of the Devonian, Laurentia, Baltica and all the bits of Avalonia had become united into a single continent. The collision zone was marked by a high mountain belt, whose eroded remnant can be seen today in Norway, the Scottish Highlands and the northern Appalachians. Their origin is attested by deformation, metamorphism, and intrusion of granites produced by melting due to crustal thickening (see Chapter 04). This mountain building event or **orogeny**, is widely known as the Caledonian Orogeny, though in the Appalachians it is usually called the Acadian Orogeny. Erosion in the Caledonian mountains shed sediment into rivers that deposited large amounts of sand in fresh-water basins in parts of the British Isles and eastern North America. These Devonian-age sandstones have long been known as the Old Red Sandstone, and the united Laurentia–Baltica continent is sometimes referred to as the Old Red Sandstone continent.

Battered though the Old Red Sandstone continent was along the multiple collision zone, its troubles were not yet over. In Carboniferous times the enormous continent of Gondwana hit it from the south, throwing up the fold mountain belt of the

Ouachitas, reactivating and enhancing many of the Appalachian structures, and deforming much of western Europe and the north-west fringe of Africa. This orogeny is referred to as the Alleghenian Orogeny in America, and as the Hercynian or Variscan Orogeny in Europe and Africa. North of the new mountain belt, the deltas of rivers from the, by then, partly denuded Caledonian mountains provided the setting for widespread coal formation. To the south, Africa, South America, India, Australia and Antarctica share glacial deposits of late Carboniferous and Permian age that attest both to their high latitude and the fact that these components of Gondwana were still united at that time.

In the Permian, Siberia collided with Baltica along the line of the present-day Ural mountains, which are a result of the associated mountain-building event. With the arrival of the rest of non-Gondwanan Asia shortly afterwards, virtually all the world's continental crust was united into a single 'supercontinent' known as Pangea (Figure 13.5). Europe (including the British Isles) and eastern North America now found themselves in low northerly latitudes. On land, desert conditions prevailed, as they do today at similar latitudes. Where the continental crust was flooded by shallow seas these were warm and poorly connected to the world ocean, so marine deposition was mostly in the form of evaporite deposits. Meanwhile volcanic activity above subduction zones and collisions with island arcs continued to extend the continent in the present-day south-west of North America.

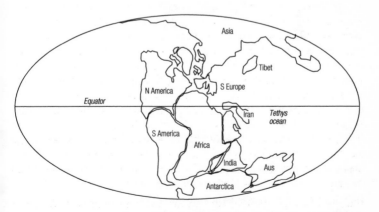

figure 13.5 The world in Triassic times (about 220 million years ago), when the continents were united into the single supercontinent of Pangea. Tethys is the name given to the wedge-shaped ocean between the northern and southern parts of eastern Pangea. (Modified from various sources.)

This was probably not the only time that all the continental crust was gathered together. It has been suggested that the situation illustrated in Figure 13.3 is a result of the break-up of a previous supercontinent that had been centred on Laurentia.

The break-up of the supercontinent of Pangea by continental rifting was as long and complicated as its construction had been. Much of the ocean floor that was created by these events still remains, especially in the Atlantic, Indian and Antarctic oceans, so we can use the pattern of sea-floor magnetic stripes (Chapter 04) to track the relative movements of the continents. Break-up was initiated late in the Triassic, when rifting began between North America and Africa, but this did not progress far enough to generate oceanic crust until the Jurassic. Shortly afterwards, India broke free from between Africa and Australia and began to be carried northwards because of subduction beneath central Asia and sea-floor spreading to its south. Next, South America split from Africa, giving birth to the south Atlantic (Figure 13.6), but the north Atlantic between Greenland and Britain did not start to open until about 65 million years ago, after a false start had attempted to split Greenland away from North America. Australia broke away from Antarctica at about the same time.

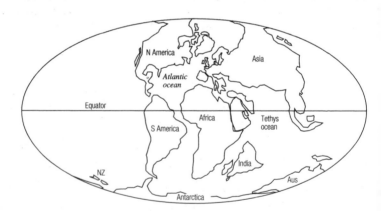

figure 13.6 The world in late Cretaceous times (about 80 million years ago). The Atlantic had been opening between North America and Africa for about 100 million years already, and between South America and Africa for about 30 million years, but had not yet begun to open between North America and Europe. India was beginning its rapid movement away from Africa and Antarctica that would lead to its collision with Asia about 40 million years later. (Modified from various sources.)

The lines along which Pangea fragmented appear to have been controlled at least partly by underlying hot-spots, where plumes rising from deep in the mantle caused heating, doming and eventually rifting of the crust. Prior to rifting, extensive fields of basaltic lava flows were erupted in some places, in the form of continental flood basalts (Chapter 05).

The westward movement of the Americas as Pangea broke apart caused the oceanic plate of the Pacific to be subducted beneath them. This is the origin of the present Andes to Alaska volcanic belt, and helped to build the Central America land bridge that currently links the two continents. As already noted, the history of western North America in particular is especially complicated because of the large number of exotic terranes that were accreted onto it.

Climate change

You have already come across several examples of climate change in this book. When trying to use geology to interpret the past, it is important to realize that some variations in climate are local, related perhaps to the changing latitude of a drifting continent, or proximity to an ocean or sea, whereas others affect the whole globe. Some of the latter are relatively brief catastrophes of the kinds leading to mass extinctions that you met in Chapter 12.

Here we will summarize the main long-term global climate changes that can be deduced from the geological record. This is essentially a history of alternation between ice ages and warm periods. The globe appears to have been particularly cold during the late Proterozoic, with glacial features found even at low latitudes during most of the 250 million years leading up to the Cambrian. The global climate then warmed up for most of the Cambrian and Ordovician, followed by a 20-million-year ice age in the late Ordovician and early Silurian when glacial deposits are found in northern Africa (which was then at the south pole) and adjacent parts of Gondwana.

Warm conditions then returned for nearly a 100 million years until an 80-million-year ice age lasting from the early Carboniferous to the late Permian. Alternate advances and retreats of the ice (glaciations and interglacials) and associated falls and rises in sea level are probably responsible for the cyclic nature of sedimentation related to the formation of coal deposits in late Carboniferous times, though undoubtedly subsidence of deltas and switching of delta channels played a role too (Chapter 09).

The world was mostly warm again from the late Permian until about the Eocene, when the climate began to cool. Antarctica was ice-covered for most of this time, but ice sheets did not spread from Greenland across North America and Europe until about 2 million years ago. As noted in Chapter 08, from the geological perspective there is no reason to suspect that our present mild climate represents anything other than a brief interglacial interlude between successive advances of the ice.

Sea level has fluctuated in various ways over geological time. Naturally, it has tended to be low at times when large amounts of water were held on land in the form of major ice sheets, and highest when there have been no major continental ice sheets. However, an even more important factor seems to be the vigour of sea-floor spreading. When sea-floor spreading is particularly active, the total volume of constructive plate boundaries in the form of mid-ocean ridges is abnormally great. This displaces seawater unusually far over the edges of the continents, giving rise to atypically extensive shallow seas.

Sea level appears to have been as much as 600 m higher than at present during the Ordovician, when sea-floor spreading following the break-up of the late Proterozoic supercontinent reached a crescendo. It had fallen to approximately its present level by the Triassic, when Pangea was in existence, but rose to a new high during the Cretaceous in response to the many young, hot, mid-ocean ridges initiated by Pangea's break-up.

Another factor that may affect the climate on a global scale is the distribution of the continents. For example in the Triassic (Figure 13.5) there was a single continental mass extending from pole to pole, which would have prevented the establishment of globe-encircling ocean currents. This is very different from the Devonian (Figure 13.4) when there was probably a current girdling the globe just south of the Equator, or to today, when a circumpolar current is able to circle Antarctica between 50 and 60°S. Another characteristic of today's climate is the monsoon, which brings much-needed seasonal rain to India and east Africa. According to computer-assisted climate models, without the barrier to atmospheric circulation posed by the 5-km-high plateau of Tibet, which is a result of the India–Asia collision, there would be no monsoon. Therefore, if there were no Tibet, the climate of the Indian ocean region, if not the globe, would be radically different.

Human-induced climate change

The greenhouse effect is perfectly natural (and without it the world would be about 30 °C colder), but fears are growing that human release of additional carbon dioxide into the atmosphere by burning fossil fuels is responsible for slight, but perceptible, global warming (and resulting rise in sea level) since about 1950. We may thus be forcing the climate to change at faster rates than most geological processes, hence mounting concern over the rate at which fossil fuels are being burned and the newly-coined term 'carbon footprint' to reflect the amount of carbon dioxide emitted by a particular activity or process. Some western countries such as the United Kingdom and Germany have managed to reduce their carbon dioxide emissions since 1990. However, the government of the USA, the country that emits by far the greatest per capita amount of carbon dioxide, has hitherto refused to adhere to reduction targets. In 2007 China, the planet's most populous country, opening new coal-fired power-stations at a rate of two per week, overtook the USA as the globe's greatest source of carbon dioxide.

The future history of the Earth

Irrespective of what may happen to climate, we can extrapolate the major aspects of present-day plate motions with reasonable confidence for the next 50 or 100 million years. It seems likely that the Atlantic will continue to widen while sea-floor spreading on the Mid-Atlantic Ridge carries the Americas away from Europe and Africa. Eventually, a major subduction zone must form in the Atlantic, but we cannot predict on which side this will happen. Much of California seems set to continue its north-westwards displacement by sideways slip along the San Andreas fault system, and is due to pass British Columbia in about 50 million years time.

The collision of India with Asia is probably a spent force, and eventually erosion will wear down the Himalayas. However, they will be surpassed in grandeur by mountains thrown up by the forthcoming collision between Australia and south-east Asia. We are witnessing its first stages in the form of thrust movements in Papua New Guinea, and about 30 million years from now the whole of the Indonesia region is likely to be compressed into a continental collision zone. At the same time, slow northward convergence of Africa towards Europe is likely to unite the two continents, replacing the Mediterranean Sea with a mountain belt higher and longer than the Alps.

Sea-floor spreading in the Red Sea will widen the gap between Africa and Arabia, and may propagate along a transcurrent fault-line that runs along the present Jordan valley. Similarly, crustal stretching across the African Rift Valley may develop into sea-floor spreading, in which case a strip from Somalia to northern Mozambique will become separated from the main part of Africa by a new ocean.

Bounded on east, north and west by subduction zones, the Pacific Ocean will continue to be ringed by Andes-type volcanic belts and island arcs. If the pace of sea-floor spreading is unable to keep up with the rates of subduction at the destructive plate boundaries, the Pacific Ocean must contract. In this case, North America will eventually swing into mainland Asia, sweeping Japan to destruction in the process, and South America will collide with New Zealand and Australia.

If this happens, then, with the possible exception of Antarctica, the world in maybe 200 million years will once again bear a single supercontinent, though with its components differently arranged from their arrangement in Pangea. Some tens of millions of years later, this too will break apart and the cycle will begin again.

Looking even further ahead, the Earth's radiogenic heat engine is running down so slowly that there is no reason to expect that plate tectonics will ever cease. The lithosphere may become a bit thicker, and sea-floor spreading a bit slower, but the large-scale processes seem likely to continue much as they do today until (according to well-established astronomical models) the Sun swells up and engulfs the Earth some 5 billion years into the future.

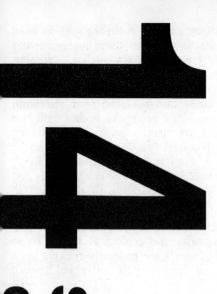

14

Solar System geology

In this chapter you will learn:
- about the geological processes that have left their marks on the surfaces of the other planetary bodies in our Solar System.

One of the benefits of studying geology is that when you have begun to understand the Earth you can use your skills and knowledge to deduce what has shaped the surfaces of other planets. We must exclude the giant planets Jupiter, Saturn, Uranus and Neptune, because these are fluid all the way down to their tremendously deep cores. However, the other 'terrestrial planets', Mercury, Venus and Mars, are rocky bodies with (we presume) iron-rich cores that are directly comparable with the Earth. As is the Moon, which counts as a terrestrial planet so far as geologists are concerned, even though it orbits the Earth rather than going directly around the Sun. The sizes, masses and densities of these bodies were compared with the Earth in Table 2.1.

There are many other bodies in our Solar System whose size and composition make them suitable for geological study. These include the larger asteroids (which are generally rocky), Io (a silicate-dominated satellite of Jupiter with active volcanoes), and at least the 15 largest icy satellites of the giant planets. Those have an icy mantle overlying a silicate core, but this ice behaves in all important respects like the silicate material that constitutes the Earth's outer layers. There are also numerous icy dwarf planets in the 'Kuiper Belt' beyond Neptune, some exceeding 2000 km in diameter. Pluto is one of these, but it is no longer the largest known example, and in 2006 the International Astronomical Union removed its status as a planet that it had held since its discovery in 1930. The first mission to Pluto will not reach it until 2015, so here I will try just to give a taste of the geology of our nearer neighbours, especially the ones that have already begun to be explored.

However, first it is worth saying something about impact cratering. This is the clearest example of a process whose effects are studied more easily on other bodies, and whose study has led to better insights into the history of the Earth.

Impact craters

Any solid surface in the Solar System that is not wiped clean by erosion, deformed out of recognition by tectonic processes, or buried (by sediments, lava or volcanic ash) becomes pockmarked with circular craters within a few hundred million years. This is because of infrequent collisions with small chunks of rocky or icy debris (asteroids and comets) that are remnants of much denser swarms of debris left over from the birth of the

Solar System. As you read in the previous chapter, the rate of bombardment fell dramatically about 3.8 billion years ago, but it still goes on. The Earth's active geological processes are fast enough to outpace the rate of crater formation, and until the 1960s most geologists did not even accept that any craters on the Earth are formed by impacts. However, as you read in Chapter 13, nearly 200 impact craters have now been identified here. In contrast, impact craters, many of them hundreds of kilometres in size, survive everywhere on the geologically less-active Moon. Its largest craters can be seen through simple binoculars. Counting the density of craters on planetary surfaces is the best way we have of estimating surface ages, because (with the exception of some areas of the Moon) we lack samples from known locations for radiometric dating.

Comparative studies of craters on different bodies, and simulations of crater formation in the laboratory and on computers, have given us a very clear idea of how craters form. Figure 14.1 shows the accepted model. A projectile (such as a fragment of an asteroid) hits the surface at about 10 to 40 km per second. At such a high velocity, the crater-forming process is driven by shock waves generated at the point of impact. There is no need for the projectile to hit the surface at right angles for the crater to be circular. The shock waves excavate a hole that ends up about 30 times the diameter of the projectile, and in doing so fling out a 'curtain' of fragmented ejecta that is dispersed over the surrounding terrain. Impact craters have a distinctive and virtually unmistakable morphology, which includes a floor lower than the surrounding terrain, a circular outline, a surrounding blanket of ejecta and (in craters above a minimum diameter that depends on the planet's gravity) a central peak.

Studies of the rock below the floor of impact craters on the Earth confirm the high pressures associated with the shock of impact. The first such crater to be thoroughly investigated was the Barringer crater, otherwise known as Meteor Crater, in Arizona. This is a small crater (barely a kilometre across) that was formed about 20 to 30 000 years ago, but it has a characteristic ejecta blanket around it, and drill holes show that the rock below the crater floor is severely shocked.

Tremendously high impact pressures are also demonstrated at the 15-million-year-old 24-km Ries crater near Nördlingen in Bavaria, Germany. This is strewn with diamonds thought to have been produced by shock compression of atmospheric

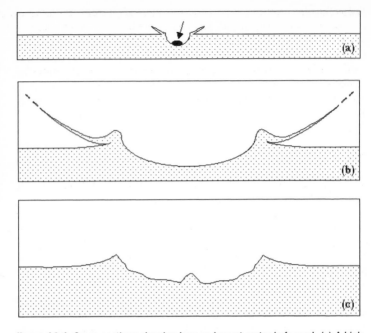

figure 14.1 Cross-sections showing how an impact crater is formed. (a) A high velocity projectile strikes the surface. A shock wave begins to excavate a crater, and also destroys most of the projectile. (b) Excavation by the shock wave spreads radially out from the point of impact. As drawn, at this stage the crater has reached virtually its final diameter, and the latest ejecta is barely able to flop over the rim. (c) A central peak has risen by rebound, and the inner walls of the crater have slumped, giving the crater its final form. Where there is no atmosphere to shield the surface, impact cratering occurs at all scales from microscopic to 1000 km. A single central peak, as shown here, occurs only for a limited range (15–140 km on the Moon, 7–70 km on the Earth). It takes about 100 seconds to excavate a 100-km crater.

carbon dioxide. Unfortunately for the good people of Nördlingen, these diamonds are microscopic in size and of no conceivable use. However, as you saw in Chapter 09, meteorite impacts have sometimes provided genuine bounty from the skies as exemplified by the Sudbury nickel ore body in Ontario.

The dangers posed by impacts are most clearly demonstrated by the 'smoking gun' of the Chicxulub crater in Mexico whose formation led to the end-Cretaceous (K-T boundary) mass extinction (Chapter 12). Even an impact as small as the one that produced the Barringer crater would destroy a nearby city, and if it hit the ocean it could displace sufficient water to cause a devastating tsunami. Telescopic sky surveys to quantify the threat posed by 'near-Earth asteroids' suggest that these are sufficiently numerous that we should expect a 100 m diameter object to hit the Earth on average about every thousand years, a 200 m object every 5000 years, and a 1 km impactor every 100 000 years.

The Moon

The Moon's diameter is 25 per cent, its mass 1.2 per cent and its density 60 per cent those of the Earth. It is too small to retain an atmosphere. Because it is so close to us, the Moon has been explored more fully than any other body beyond the Earth. A calibrated cratering timescale has been established using the relationship between radiometric ages of samples of lunar material and the local density of impact craters. By extrapolating this, we can make a reasonable estimate of the absolute ages of cratered regions on other bodies in the inner Solar System, though this does not apply beyond the orbit of Mars where the impact history has been too different.

The Moon appears to have cooled down a lot quicker than the Earth. This is to be expected, because a planetary body generates heat from radioactive decay in proportion to its volume but loses heat in proportion to its surface area. The smaller the body, the more surface area it has relative to its volume, so the faster it will cool. In the case of the Moon, this cooling has resulted in a present-day lithosphere about 1000 km thick, which explains why the lunar surface shows no clear signs of current or recent geological activity.

The Moon's geological processes waned so quickly that, as noted in the previous chapter, large regions of primary crust have survived in the form of the pale, anorthosite highland terrain, dating back about 4.4 billion years. The dark areas on the Moon are covered by basaltic lavas that flooded most of the largest impact basins. This happened mostly between about 3.9 and 3.1 billion years ago (dated radiometrically by samples collected during the 1969–72 Apollo Moon landings), though the youngest flows may be as 'young' as only 2 billion years. These dark areas were once (incorrectly) thought to be lunar seas, and they are still known by the Latin word for sea 'mare' (pronounced MAH-ray; the plural form is 'maria' pronounced

figure 14.2 A 200-km-wide view of the Moon, photographed from orbit by Apollo 15. Solar illumination is coming from the right. Rugged highland crust (in this example uplifted as part of the rim of an enormous impact basin) occupies most of the view, but is flooded in the north-west by smoother mare basalts. The sinuous trench is a collapsed lava tube, wider than those on Earth because of the Moon's lower gravity. There are more craters in the highlands than on the mare surface, but none in this view is large enough to have developed a central peak.

MAH-ria). A rapid decline in the rate of bombardment in the early Solar System is demonstrated by the fact that the lunar highlands typically have far more craters per unit area than the maria, despite being only about 25 per cent older. Primary crust partly flooded by mare basalts can be seen in Figure 14.2.

The craters in the lunar highlands show virtually no signs of deformation by tectonic processes, so the Moon's lithosphere must have become thicker (and/or more rigid) than the Earth's at an early stage. After the most recent episode of mare lava eruption, the lithosphere evidently became too thick for any more magma to escape from below. Except for impact cratering, the Moon has been geologically very quiet ever since.

Seismometers left in place by the Apollo programme detected weak 'moonquakes' (some of them near the base of the Moon's thick lithosphere), and there are occasional controversial reports of gases leaking from isolated patches of the surface.

Mercury

Mercury is the closest planet to the Sun. Though larger than the Moon, it is still too small to retain an appreciable atmosphere. Although its density is 98 per cent that of the Earth, this is actually anomalous because Mercury has only a twentieth of the Earth's mass and so lacks the gravity needed for the same amount of 'self compression' in its interior. Mercury's high density in comparison to its mass can be reasonably explained only if it has a much larger iron core, relative to its total size, than the Earth, or indeed than Venus or Mars. If so, Mercury must be made of a greater proportion of iron and a smaller

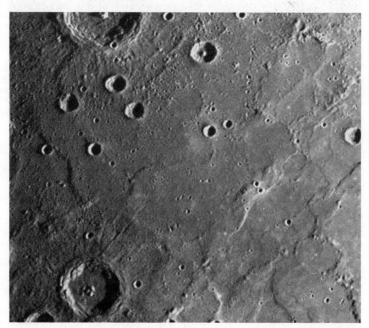

figure 14.3 A 600-km-wide view of Mercury. Solar illumination is coming from the right. The larger craters have obvious central peaks. The wrinkles on the surface probably indicate a lava surface, but some may be a result of tectonic shortening.

proportion of silicates than any other terrestrial planet. How it came to be made from such a different recipe is such an intriguing mystery that both NASA and the European Space Agency are sending probes there, the first since NASA's Mariner-10 sent back images in 1974–5. Those revealed Mercury as a heavily cratered world, superficially reminiscent of the Moon but lacking its dark flood basalts. Mercury appears to have been geologically dead for most of the age of the Solar System, but signs of ancient volcanic and tectonic events can still be seen (Figure 14.3).

Venus

In terms of basic properties, Venus is our twin planet. Its diameter is 95 per cent, its mass 82 per cent and its density 95 per cent those of the Earth. It is shrouded beneath an opaque atmosphere, but orbiting spaceprobes have been able to image the surface using radar. Venus's atmosphere is denser than ours, with a surface pressure over 90 times greater, and is 96 per cent carbon dioxide. Venus's average surface temperature of 450°C is much hotter than expected for a world whose distance from the Sun is only a third less than the Earth's.

How Venus got this way is a tale of extreme climate change. Venus probably once had liquid water in similar abundance to the Earth, but the higher temperature closer to the Sun meant that a greater proportion of it was evaporated into the atmosphere. Water vapour acts as a greenhouse gas, so the atmosphere became even warmer, so even more water evaporated, leading to a 'runaway greenhouse effect' ending up with the hot dry world we see now. Venus's atmosphere today contains only a trace of its previous water vapour, because solar ultraviolet radiation has long since split most of it into hydrogen (which has escaped into space) and oxygen (which has reacted with the rock). Venus's present greenhouse effect is maintained by the large amount of carbon dioxide in its atmosphere. Earth probably once had a similar amount of gaseous carbon dioxide, but most of this is now locked up in limestone, a rock type that can form only in water and so is not expected on Venus.

The surface of Venus is better documented than much of Earth's ocean floor, because virtually all of it has been mapped by radar images in sufficient detail to show features as small as about 100 metres across. However, our understanding of Venus does not compare with that of the Earth, because we lack seismic data and other information that can only be gathered on the ground.

Perhaps the best way to sum up Venus's geology is to remark that it does not appear to be experiencing Earth-style plate tectonics. There are widespread volcanoes (Figure 14.4) associated with lava flows that appear to be basalt, and a few highly deformed regions, but nobody has come up with a convincing model to link the various elements together on a global scale in the way that plate tectonics does for the Earth. A likely explanation is that the dryness of Venus's interior means that the top of its asthenosphere is not weak enough to allow plates to glide freely across it, in contrast to the Earth where ocean water is continually fed into the asthenosphere via subduction zones.

The age of Venus's surface is particularly perplexing. Crater counting suggests an average age of about 500–800 million years. It was once claimed that there is no significant age variation from region to region, which forced the conclusion that the whole globe had been resurfaced in a 10-million-year orgy of flood basalt volcanism about 500–800 million years

figure 14.4 A perspective view across part of Venus, constructed from radar images obtained by the Magellan probe in orbit above the planet. The scale varies from about 100 km across the foreground to nearly 1000 km across the background. The nearby fractured plains (dark) are overlain by lava flows (pale) emanating from the shield volcano Sapas Mons in the middle distance. Other volcanoes dominate the horizon.

ago. Now it seems more likely that the rate of resurfacing has been on the wane ever since about 800 million years ago and that there are probably small tracts of terrain as young as a few millions of years old. Certainly some of Venus's volcanoes look young, though there is no proof of present-day eruptions. The few (possibly unrepresentative) images we have from the surface show it to be lava strewn (Figure 14.5).

figure 14.5 The surface of Venus, seen from the Soviet lander Venera 13. Crude chemical analysis of the slabby rocks beside the foot of the spacecraft show a basaltic composition.

Mars

Mars is smaller than Venus, but larger than Mercury. Its diameter is 53 per cent, its mass 11 per cent and its density 72 per cent those of the Earth. Like Venus, its atmosphere is mostly carbon dioxide, but the surface pressure is less than a hundredth of the Earth's.

Mars appears to have the cool, thick lithosphere that we would expect for a planet of such small size. It has several impressive volcanoes, including the largest in the Solar System. This is

Olympus Mons, whose summit is some 24 km above its base. It would dwarf the Big Island of Hawaii, which is the Earth's largest volcanic edifice (Figure 14.6). The vast bulk of Olympus Mons can be supported only by a thick, strong lithosphere. Moreover, the concentration of volcanic activity at this point over many hundreds of millions of years suggests that Mars's lithosphere is stationary relative to the hot plume within the mantle that presumably feeds the volcano.

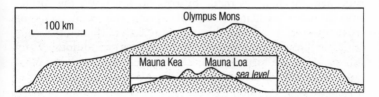

figure 14.6 Cross-sections to show the comparative sizes of Olympus Mons on Mars and the Big Island of Hawaii (its two largest volcanoes are named). The horizontal scale is as shown, but the vertical scale is exaggerated fivefold.

Large though it is, Olympus Mons covers too small an area for impact crater statistics to give a reliable estimate of its age. The best guess is that it last erupted about 30 million years ago. There are several smaller (but still very large) volcanoes nearby and others elsewhere on the globe.

For a smallish planet, Mars has a surprising number of exceptional features in addition to its volcanoes. The most fundamental is a two-fold division into an ancient, highly cratered highland terrain occupying most of the southern hemisphere and a younger low-lying terrain in the north. The highland terrain has many large channels upon it, some of them in dendritic networks (e.g. Figure 8.4a), which can hardly have been cut by anything other than flowing water and would appear to have been fed by rainfall. The younger surface age of the northern lowlands is probably due to deposition of sediment transported from the highlands. It has even been suggested that part of the northern hemisphere was once occupied by a shallow ocean. That being so, it would be rather surprising if primitive life had not developed, and searching the conditions necessary to sustain life and ultimately for life itself are prime goals of missions to Mars.

Liquid water would not be stable at the surface under the low atmospheric pressure that prevails on Mars today, so clearly its climate must have been very different when water was flowing in the large channel networks. Mars has lost much of its atmosphere and cooled down over time, in contrast to Venus, which has warmed up and developed a dense atmosphere.

Based on ages inferred from crater counting, the 'wet' period of Mars's history had ended by about 2 billion years ago. Images and chemical analyses made by robotic rovers on the surface have found convincing proof of Mars's wet past, including cross-bedding produced by flowing water (Plate 14), and minerals that can only form in hydrous (wet) conditions. Water still occurs in the form of ice in the northern and southern polar caps. In winter these are enlarged by a seasonal covering of frozen carbon dioxide. There is at least one example of what may be a dust-covered frozen sea (Figure 14.7) and a large amount of water may also be stored as ice within the martian subsoil. High-resolution images from orbiting spacecraft have revealed gulleys in crater walls, some of which appear to be experiencing brief episodes of flow even today (Figure 14.8). The mechanism for melting and releasing this water from the subsoil is unknown. Even the longest young gulley is only a few kilometres in length, so they are on a far smaller scale than the ancient dendritic networks, and most of the water flowing down them must evaporate into Mars's atmosphere before it reaches the bottom.

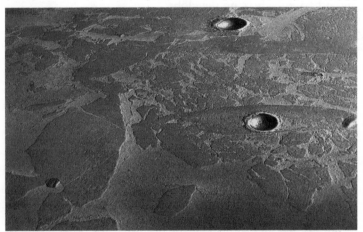

figure 14.7 This 40-km-wide view of Mars, from the European Space Agency's Mars Express orbiter, may show the dust-covered remains of a frozen sea.

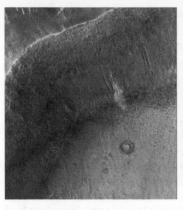
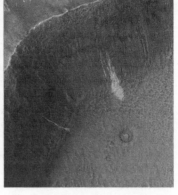

figure 14.8 Views of part of the inner wall of the same martian crater in 2001 (left) and 2005 (right). Several narrow gulleys can be seen. A new pale deposit in a gully in the lower left of the 2005 view suggests that water flowed in this gulley between the two dates.

With the exception of these local and ephemeral discharges of water, sediment transport and erosion of surface features are today restricted to landslides and wind effects. Extensive fields of sand dunes occur, and dust storms sometimes obscure the surface for weeks on end.

Mars is also notable for Valles Marineris, a 4000-km-long canyon system in the southern highlands that reaches 7 km in depth and over 600 km in width. This would dwarf Earth's Grand Canyon even more than Olympus Mons would overshadow Hawaii. It probably owes its origin to tectonic fracturing, but has been extensively modified by flowing water and landslides.

The asteroids

As noted in Chapter 02, the asteroids, mostly orbiting the Sun between Mars and Jupiter, are probably the remnants of planetesimals prevented from accreting into a planet because of gravitational perturbations caused by Jupiter. Most are rocky bodies, though some appear to be metallic (like the Earth's core). The largest of them, Ceres, is about 900 km across, which is far too small for radiogenic heat production to keep its interior warm enough to be mobile today. All the asteroids have thus long-since become lithospheric throughout, and are not expected to display evidence of recent internally-driven geological processes.

However, images of Vesta (the third largest asteroid, 500 km in diameter) obtained by the Hubble Space Telescope are clear enough to reveal large dark patches on its surface. These could be a dark interior, exposed at the surface as a result of collisions. If so Vesta would seem to have a compositionally differentiated structure, analogous to the crust-mantle distinction in the Earth. Alternatively, the dark patches could be lava flows. Either interpretation calls for Vesta's interior to have been hot at one stage. If this is correct, then the explanation probably lies with short-lived radioactive isotopes, such as aluminium-26 (^{26}Al), which we have good reason to believe could have been very significant heat sources during the Solar System's first hundred million years or so. We may find out when NASA's 'Dawn' mission arrives at Vesta in 2011.

Several of the smaller asteroids have now been imaged close-up by passing spacecraft. These are confirmed to be irregular in shape, but there is considerable variation in the abundance of impact craters. Some have curious dust-filled depressions, and a surprising number of boulders strewn across their surface (Figure 14.9).

figure 14.9 An oblique view across a 1.4-km-wide part of the asteroid Eros, seen from a range of 50 km in 2000 by NASA's NEAR-Shoemaker probe. Impact craters can be seen, but also boulders up to 60 m in size.

Io

Io is the innermost of Jupiter's four large satellites. By analogy with the Moon, which it resembles in size, density and inferred composition (except for having a larger core), we would expect Io to be a densely cratered body with a few signs of geological activity in the past 2 or 3 billion years.

The reality is very different, because Io is actually the most volcanically active body in the Solar System. Up to about ten volcanoes are erupting at any one time (Plate 15), and the surface is being covered by volcanic material at a rate more than adequate to bury impact craters as fast as they are formed. There are both explosive eruptions, in the form of 300-km-high eruption plumes, and lava flows. When first seen, on images from the Voyager spaceprobes in 1979, the lava flows were widely interpreted as sulphur because of the red or yellow colour of Io's surface. Subsequent thermal infrared studies demonstrated that the temperatures in the volcanic vents are too high for molten sulphur, so it seems that we are dealing with Earth-style silicate lava (basalt or an ultrabasic lava). Io's colour is almost certainly caused by sulphur, but it is merely a surface dusting, as is sometimes seen on lava flows on the Earth.

Radiogenic heating cannot account for Io's prodigious volcanism. The explanation appears to be tidal heating of Io's interior, caused by tidal deformation of Io's shape as it orbits Jupiter.

Europa

Europa circles Jupiter outside the orbit of Io, and completes one orbit in exactly the time taken for Io to complete two orbits. This relationship is responsible for maintaining the rate of tidal heating in the two bodies, which is weaker in Europa than in Io. Europa is slightly smaller and has an icy surface. However, its average density is not much less than Io, and so the ice can be no more than about 100–150 km thick above its silicate crust and mantle. Close-up images of Europa's surface show good evidence that only the outer few km of ice is solid (Figure 14.10), below which there is likely to be a global ocean, and below that maybe a less-active, submarine, version of Io.

figure 14.10 A 35-km-wide region of Europa. The icy surface, which has a characteristic ridge-and-groove pattern, is relatively intact in the north-west but has begun to fracture. The further south-east you look, the more fractured the ice becomes, until all you see is isolated rafts of ice dispersed across a now re-frozen sea.

If there are zones of hot rock not too far below the ocean floor, then surely there is hydrothermal circulation and chemical reaction between rock and water, and thus a likelihood of hot springs on the ocean floor like the 'black smokers' that you first met in Chapter 05. As you read in Chapter 13, this is the setting where life probably began on Earth, so why not on Europa as well? Missions to Jupiter are both difficult and expensive, but such is the lure of finding life that several space agencies are planning experiments that can be done from orbit and ultimately on Europa's surface to prove the existence of the ocean, and to find ways to sample it – by landing beside tidal cracks where slush from below is squeezed up, or even by drilling through the ice to allow a robotic submarine to go exploring.

Large icy bodies

There are several other icy satellites that should interest the geologist. The next one out from Jupiter is Ganymede (Figure 14.11), which is bigger than the planet Mercury. However, its density is much less because it consists of about three parts ice and one part rock (Figure 14.12). This preponderance of ice over rock is a common factor from now on. Jupiter has a fourth large satellite, Callisto, beyond Ganymede. Saturn has one large icy satellite, Titan, that rivals Ganymede in size, and six others more than 400 km across, Uranus has five satellites larger than this, Neptune has two, of which Triton (2700 km across) is largest and most interesting. The Solar System's inventory of large icy bodies is completed by the largest of the icy dwarf planets in the Kuiper belt beyond Neptune, such as Pluto (2300 km diameter), its most substantial satellite Charon (1250 km diameter), and Erys (2400 km in diameter) that was discovered as recently as 2003.

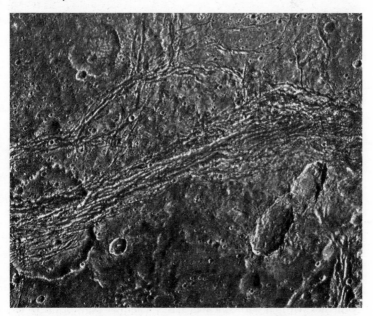

figure 14.11 A 110-km-wide area on Ganymede. The large impact crater in the south-west has been rifted apart by faults associated with north–south extension. The strange elongated crater in the south-east was probably formed by the impact of a comet that had broken into at least two fragments before it struck the surface.

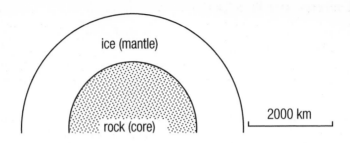

figure 14.12 Cross-section through a large icy body, showing differentiation into a dense rocky core surrounded by less dense ice. The scale bar is for Ganymede (the largest example of its class), but many smaller icy satellites and bodies such as Pluto probably have similar structure. Ice is not necessarily just frozen water; it can include other 'volatile' substances such as methane, ammonia, carbon dioxide and nitrogen.

We can ignore the hundred or so other known satellites of the outer planets that are less than about 400 km across, and a huge number of Kuiper belt objects below the same size threshold, because below this size the body's own gravity is too weak to pull it into a spherical shape. Unlike some of the asteroids, we have no reason to suspect internal differentiation or other geological processes in such irregular bodies.

It may seem strange to deal with predominantly icy bodies in a book about geology, but their relevance becomes clear if we review the properties of ice. At the low surface temperatures prevailing in the outer Solar System (–150 °C at Jupiter and –240 °C at Neptune) ice is extremely strong and rigid, just like rock on the Earth. However, in the interior of an icy body with a supply of heat, ice is capable of flowing by convection without melting. This distinction between rigid and convecting behaviour is exactly the distinction described in Chapter 02 between the lithosphere and asthenosphere of the Earth (or, indeed, any other terrestrial planet), except that we are dealing with ice rather than silicates.

A further parallel with silicates arises when we consider how ice melts. Unless it is pure, ice does not melt at a single temperature. Salty ice, such as we might expect in Jupiter's icy satellites because of reactions between water and rock, begins to melt at a few degrees below 0 °C, sweating out a brine instead of pure water. Ice that consists of a mixture of frozen water and frozen

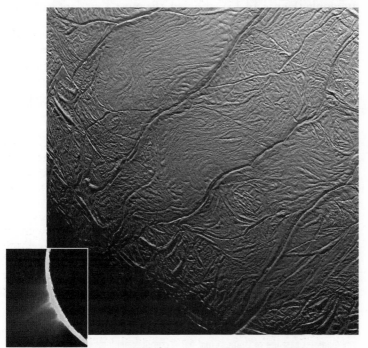

ammonia, as is likely at Uranus and beyond, begins to melt at about −100 °C, yielding a liquid that is about two parts water and one part ammonia. Both are examples of partial melting, just like what happens in the mixtures of silicate minerals that make up rock. 'Ice' becomes even more like mineral-rich silicate rock at Neptune, where methane and carbon dioxide are added into the mix.

Thus, with a lithosphere/asthenosphere distinction and the potential to produce compositionally distinct 'magma' by partial melting, icy bodies have the essential attributes of a terrestrial planet. There is too little rock within them for much radiogenic heating, but fortunately for the geologist most of them show evidence for current or previous episodes of tidal heating. For example, parts of Ganymede's surface are intensely fractured, but the number of superimposed impact craters shows that the fracture pattern was created in the distant past.

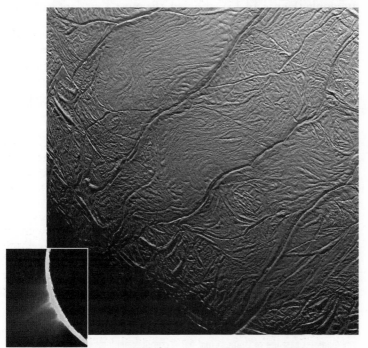

figure 14.13 Main view: a 120-km-wide region of Enceladus, revealing the intensely fractured nature of its crust in this region. The largest fractures are warm, and are the source of ice particle eruptions. Inset: a reduced-scale view of part of Enceladus, seen looking towards the Sun, so that only a narrow crescent is illuminated. Jets of ice particles erupted into space from the region shown in the main view are revealed by the way they scatter the sunlight.

There is not space here to describe each of these fascinating bodies. Highlights include a 500-km-diameter satellite of Saturn called Enceladus, whose surface is criss-crossed by fractures. Images from a spacecraft orbiting Saturn have captured jets of ice crystals venting into space from some of these, presumably powered by tidal heating (Figure 14.13). This is the icy equivalent of pyroclastic volcanism.

Moving out to the satellites of Uranus, we find features such as 'lava' flows sketched in Figure 14.14, which are thought to be composed of an ammonia–water mixture produced by partial melting. It is known that such a mixture is vastly more viscous than pure water, and would behave very much like a silicate lava flow on Earth. Neptune's large satellite Triton has a seasonal polar cap of frozen nitrogen, punctured by geysers powered by pressurized escape of nitrogen gas, and a surface geology dominated by other ices including methane, carbon dioxide and ammonia (Figure 14.15), although the most abundant ice in the mantle is probably water.

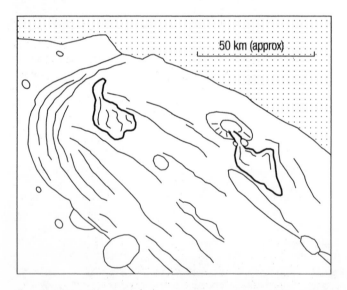

figure 14.14 Sketch made from an oblique view of part of Miranda (a 470-km-diameter satellite of Uranus) that has a complex geological history. Several impact craters are visible (roughly circular). One 10-km crater in the lower left has been partly obliterated by a resurfacing event, whereas another is superimposed over the boundary and so must post-date the resurfacing. Two icy 'lava' flows are shown in bold outline.

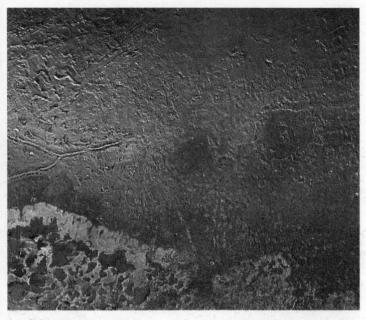

figure 14.15 A 600-km-wide region of Triton. The ragged fringe of the solar polar cap of nitrogen ice runs across the bottom. Beyond this, a complex dimpled icy surface is cut by fissures and buried by patches of smoother material that may be icy lava flows.

To conclude, we should backtrack to Saturn's largest satellite, Titan. This is the most Earth-like body in the outer Solar System from the point of view of the diversity of geological processes that it exhibits. Titan has a dense smoggy atmosphere in which the most abundant gas, like the Earth, is nitrogen. The atmospheric pressure at the surface is 50 per cent greater than Earth's, but rather than oxygen, carbon dioxide and water vapour (as on the Earth) the rest of Titan's atmosphere is mostly methane and it is much colder (–180°C at the surface). The smog, which makes the surface hard to see, is made of hydrocarbon molecules created by the linking together of methane under the influence of solar ultraviolet radiation. Neither active icy volcanism nor tectonics has been proven on Titan though they seem likely, and there are certainly wind-blown dunes (made of ice grains rather than sand). However, what makes Titan so special is that it has its own equivalent of Earth's hydrologic cycle. Instead of water, this is based on methane. From time to time (possibly seasonally) methane condenses in the atmosphere to make methane rain. This feeds some impressive dendritic river networks. Low-lying areas across most of the globe are currently 'dry', but in the polar regions there are lakes and seas (presumed to be liquid methane) that are up to several hundred kilometres across (Figure 14.16). One day, a geologist from Earth may stand on the shore of one of those methane seas to study coastal erosion such as you saw at the end of Chapter 08, and sediment transport on beaches of icy sand analogous to the terrestrial beaches in Chapter 09.

figure 14.16 A 250-km-wide region of Titan, seen by the imaging radar onboard the Cassini orbiter, showing a rugged coastline with numerous offshore islands. This is a very Earth-like scene, except that the land is made of ice, and the sea is made of liquid methane.

Geology among the stars

Planets have now been discovered around more than 200 other stars. So far the discoveries have been mostly Jupiter-sized or bigger, orbiting rather close to their stars, because these are the easiest to detect. However, rocky planets comparable in size to the Earth and cool enough to have liquid water have begun to show up as well. In our lifetimes we should get spectroscopic data on their atmospheric compositions, but we cannot expect to see their surfaces in sufficient detail for any kind of geological study. However, with many unsolved questions remaining about the Earth, and the geological exploration of our Solar System barely begun, there is plenty to keep the geologist busy, as well as fascinated, nearer to home.

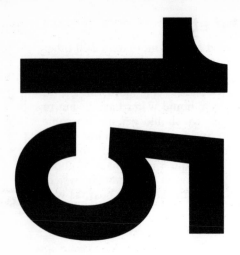

15
going into the field

In this chapter you will learn:
- about the importance of making intelligent observations in the field
- to develop a safe and responsible attitude to fieldwork.

Using spacecraft to explore the surfaces of other planetary bodies is both fun and informative. Detailed images of the Earth itself from space can also be illuminating, but there can be no denying that for many geologists, whether professionals or hobbyists, the main attraction of geology is doing it in the field. Fieldwork can be a great source of pleasure, but it is also the best way to appreciate the sort of evidence from which most of our understanding of the Earth's history is derived.

When I was doing fieldwork in the Arabian desert, a local once asked me, 'Where exactly are these fields that you work in?' Perhaps the term **fieldwork** is another of piece of geological jargon that seems designed to confuse the layperson. However, in this case at least it's a term that we have in common with botanists, zoologists, ecologists and even sociologists. When a geologist speaks of 'going into the field', he or she means spending time outdoors in a place where the rock sequence or geological process to be investigated can be seen. Thus, fieldwork can be conducted in places as diverse as a sandy beach, a roadside cutting, a remote stretch of moorland or the flanks of an active volcano.

If you belong to a geology club, or are following a course of study in geology, you have probably been on at least one field trip. However, if you have never been in the field and have decided, after reaching this stage in the book, that you want to go deeper into geology, then I urge you to get out there and begin looking at some rocks. This chapter suggests some ways to begin practical geology in the field. It therefore describes things to look out for, discusses the equipment you will find useful, and suggests a code of practice to enable you to work safely and without annoying any landowners.

Things to do in the field

It is sensible to decide in advance what you want to achieve during a field trip. For example, are you just going out for a general look at the rocks, or do you want to search for the answer to a particular riddle? If you are going in a group organized by somebody else, you can leave the decisions about where to go and what to look at to them, but to get the most out of a field trip it pays to approach each locality in a methodical fashion.

When you arrive at a rock **exposure** (a place where bedrock is visible), perhaps a cliff face, do not rush straight up to it. I will have something to say about the safety of working under cliffs shortly, but it makes scientific sense to stand back for a while anyway.

You should spend time taking in the general features of your site. These are best seen from a distance. The sort of thing to look out for is whether or not there is any layering in the rock, which might take the form of bedding, layers within an igneous intrusion, metamorphic foliation, or a set of fractures (joints). You may not be able to decide which of these four you are seeing, but for now you should be content with spotting the pattern; you will be able to distinguish between the options when you get within touching distance.

While you are still ten or more metres away, you should look for variations. For example, if you can see some layering, try to decide whether these layers maintain a uniform orientation throughout the exposure, or whether they are folded.

You should also visually trace some layers sideways to see if they are continuous. If they are not, try to decide whether this is an inherent property of the layers (perhaps each one represents a deposit laid down in a channel that had a finite width, or maybe you are seeing large-scale cross-bedding of the sort that occurs in wind-blown sand dunes), or whether originally continuous layers have been truncated by an igneous intrusion or by a fault (which will almost certainly visibly offset the layers).

If the structure within an exposure is complicated, making a simple sketch can often help you begin to sort things out. This should not be a piece of representational art, showing every crack, shadow and tonal variation. Rather, it should show only those features that are of geological significance. For example, bedding is significant and should be drawn, but if you try also to show all the irregularities in a rock face left by either natural weathering or quarrying then you will obscure the essential features. Always include a scale in a field sketch. As a rough guide, most adults are 1.6–1.8 metres tall, so if you can persuade somebody to stand against your exposure you can very easily mark a scale-bar on your sketch. A sketch is usually more useful than a photograph, because it shows only what you need and is easy to annotate while you are in the field.

Before you finish your preliminary survey, choose a suitable place to begin your close-up examination. Avoid approaching anywhere you are likely to be hit by material falling from above, or where the ground beneath you could be unstable. Furthermore, do not head straight for somewhere that looks unusual. Instead, choose a spot that looks representative of the exposure in general. It is best to try and make sense of any strange areas after you have worked out the main story.

By the time you arrive at the rock face, your preliminary observations will have suggested a number of checks needed to test any hypotheses that have begun to form in your mind. If you have decided that this exposure was probably sedimentary rock, you will expect to be able to confirm this by close inspection, which should reveal the individual grains making up the sediment (unless these are very fine), small-scale bedforms and (if you are lucky) trace fossils and/or body fossils. On the other hand, a coarse-grained igneous rock will manifest itself with clearly visible interlocking crystals, including minerals of the sort that do not survive transport in the sedimentary regime (Chapter 08). In order to identify the rock, you will need to recognize and identify its main minerals, with the aid of Appendices 1 and 2 or a more detailed field guide to rocks and minerals.

You may see features close up that are worth sketching or photographing. Examples would be small-scale cross-bedding, fossils, or the shape of the contact at the edge of an igneous intrusion. Just as with any more distant sketch (or photograph), you should try always to include an object that will give a clue as to scale: a coin, a lens cap or a finger are probably the most common scale objects. The latter has the advantage that it is unlikely to be accidentally left behind when you leave.

If you are not visiting a quarry or a cliff, you are probably dealing with comparatively small exposures, unless you are in an arid environment where there is little or no soil cover. However, the principles outlined above still apply. Stand back and assess the visible area of rock as a whole, however small, before rushing in and pressing your nose against it.

Field equipment

It is a common misconception that geologists spend their time digging up rocks and fossils. It is rare for a geologist to go into the field armed with a spade. The trick is to go somewhere where the things you want to see are already at the surface. In many cases, the natural weathering of the rock picks out features more clearly than they can be seen on freshly exposed surfaces. Exposures of rock are most common in hilly or mountainous regions, especially along water courses, but they may be particularly good on the coast. Quite apart from the damage and disturbance that digging would cause, the labour involved in removing two or three feet of topsoil will rarely be repaid by the small patch of dirty and possibly frost-shattered bedrock that this would reveal.

The most important items of field equipment are a perceptive eye and an enquiring mind. There are no artificial aids for the latter, but the former can be augmented by carrying a hand lens. A typical hand lens is smaller than a Sherlock Holmes-style magnifying glass, but provides a magnification of about ten times. This can be extremely helpful for such things as determining the shapes (angular or rounded) and textures (frosted or glassy) of sedimentary grains, spotting cleavage traces in minerals, and making out fine details in fossils.

You should also carry a notebook, for recording your observations and making sketches. Pencil is best because it can be rubbed out to correct mistakes, and does not smudge as badly as ink when it rains. If you might want to go back on another day to check or extend your observations, it is important to record your exact location. Marking this on a map and/or a sketch map of your own should do the trick. Noting down co-ordinates read from any GPS (Global Positioning System) device that you are carrying is second best.

Sometimes you may want to collect a small sample. If you are just starting out as a geologist it is a particularly good idea to build up your own collection of the main types of rocks and fossils, but see my later warning about selfish and indiscriminate collecting. If you do find something that you want to take away, then you need something to put it in, and some way of labelling it. One specimen per small plastic bag, each containing a sample number on a slip of paper is a good way to proceed, provided you record the location of each find in your notebook.

If not with spades, then it is with hammers that geologists are usually associated. Unlike spades, hammers are a standard item in many geologists' field gear. In situations where the rock surface is too badly weathered or covered by lichen for it to be identifiable, a hammer can be invaluable for breaking off a piece to expose a fresh surface. However, natural freshly broken pieces can very often be collected near the foot of cliffs, in which case there is no need to hammer the rock face at all. The same goes for collecting fossils; you will usually find many more, and avoid damage, by searching through rock falls or spoil heaps. If you do use a hammer, then it must be a purpose-made geological hammer. The head of a hammer sold for knocking in nails will not be made of adequately tempered steel, and is liable to splinter if struck against a rock. A steel shard from a shattered hammer can blind you if it hits you in the eye.

The other standard item of field equipment is a device called a compass clinometer, which is used to measure the orientation of bedding, foliation, fault planes and so on. This is particularly important if you are being ambitious enough to draw your own geological map. It is valuable too when, for example, you are visiting several isolated exposures of a sedimentary rock and want to determine if the bedding in each one has the same orientation. If it does not, then the rock sequence has probably been folded.

A compass clinometer has a magnetic needle, functioning as in any ordinary magnetic compass, and a second, non-magnetic needle that is pivoted from one end so as to hang vertically under its own weight. To appreciate how to use a compass clinometer, it is necessary to understand how the orientation of any plane surface can be specified in a unique and unambiguous manner. This can be done with just two measurements. The first is to imagine a horizontal plane intersecting the geological surface (such as a bedding plane) whose orientation we want to determine. The two planes must intersect at a horizontal line and the angle between this line and north is defined as the **strike** of our geological surface (see Figure 15.1). To measure the strike, the base-plate of the compass clinometer is held horizontally, with one edge pointing along the strike direction. The angle that this edge makes with the free-floating magnetic compass needle is the strike, and can be read on the dial. The only other measurement we need to know is how steeply the geological surface is dipping, relative to horizontal. This is called the **dip** of the surface. To determine this, the compass clinometer is held with one edge running directly down the steepest gradient that can be found on the geological surface. The dip can then be read using the free-hanging needle.

In the example shown in Figure 15.1, the strike is 150° and the dip is 35°. This would be recorded in our field notebook as 150/35. This is unambiguous if strike is always measured clockwise from north (as in this example), but to play safe it is

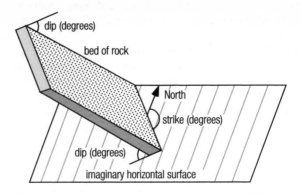

figure 15.1 Definitions of strike and dip. A bed of rock is used as the example, but the definitions apply to any geological feature (such as a dyke, a joint or a fault) whose local orientation can be expressed as a plane. Strike is the angle, relative to north, made by any horizontal line on this geological surface. Dip measures how steep the surface is, and is defined as the angle of slope measured downwards from horizontal.

figure 15.2 A compass clinometer being used to measure the strike of the edge of a pale igneous dyke, which has been intruded into a darker rock. The photograph was taken looking down onto a roughly horizontal surface. In this example, the strike is approximately 160 ° relative to magnetic north. The edge of the dyke dips at 80 ° towards the east (left in this view), so the strike and dip of the dyke would be written down as 160/80E.

common to note down the approximate direction in which the surface is dipping. In this example it is dipping down towards the north-east, so we could record 150/35NE. Figure 15.2 shows a compass clinometer in use.

If you want to record your visit photographically, it is best to have a camera with adjustable focus, or even a choice of lenses. Some landscapes are best recorded using a wide-angle lens, but detailed close-up photographs require a macro lens. Digital cameras allow you to check your photo as soon as you have taken it, but beware that the resolution of the camera display is probably too poor to confirm that fine details are in focus.

Other items of field equipment depend on the nature of the rocks. If it is in a poorly cemented sandstone, you may find a trowel or similar object handy for scraping away loose grains in order to reveal the original cross-bedding. If it is a shale and you hope to find fossils, a chisel comes in handy for splitting the rock along its bedding surfaces.

Finally, if there is any likelihood that your field trip will involve working close to steep or vertical faces exceeding about three metres high then you should make sure that you take a hard hat. These are cheap and readily obtainable from builders' merchants and do-it-yourself stores.

Field safety

In the field, you should take care to minimize the likelihood of injury to yourself and to others. As already noted, the most basic precaution is to assess the inherent risks before approaching a rock face. You should wear your hard hat whenever there is the slightest chance of anything gravel sized or above falling onto you. While it won't help you survive a major rock fall, a hard hat can save you from concussion or a worse head injury caused by small lumps of debris.

If it is absolutely necessary to use a hammer, then you must make sure that nobody is within about three metres of you. Rocks such as flint and many igneous varieties are particularly hard and brittle. Hammering these can send sharp splinters flying. Most older geologists have met someone who has lost the sight of an eye in this way, and it is foolish to hammer without wearing goggles or safety glasses.

Be adequately clad. Walking boots with ankle protection are the most sensible footwear in rough terrain, and you should carry protection against sun, rain and cold. When visiting a coastal exposure, be aware of the state of the tide and plan ahead to avoid being cut off. When visiting a remote area, general backcountry rules apply: be wary of dangers such as being caught in a gully or canyon in the event of a flash flood, make sure you have enough food and drink to survive if your return is delayed for some reason, and leave a copy of your intended route with someone in case you have to be rescued. Consider taking a mobile phone, or similar means of communication.

Fieldwork code of practice

The rules governing geological collecting and other activities vary from place to place. However, there are various widely applicable guidelines.

- If you are entering a quarry or other private land, seek permission. Keep to paths, close gates behind you, and do not climb over walls or fences. Quarry owners will usually insist that you wear a hard hat at all times, and that you stay well clear of machinery.

- If you are working in a national or state park, nature reserve, or other conservation area, then find out the rules about collecting. Some places will allow you to collect samples only from loose debris, while others ban collecting entirely.

- Even where there are no local rules, be aware of the damage you might cause. Hammer, if at all, with discretion. Do not damage rock faces, and do not try to extract fossils or mineral specimens that are well embedded. You will usually end up wrecking them, so it is much more sensible to take a photograph, and leave the item in place for others to admire.

Geological maps and guidebooks

If you want to know about the geology of a particular area, then an obvious first step is to buy a geological map. A good bookshop will be able to order these for you, and some can be obtained from internet book suppliers. A geological map usually shows the **outcrop** of each rock unit, which is distribution of rock types that would be seen at the surface if the superficial deposits (such as soil) were removed. Sedimentary rocks are usually colour-coded by age, whereas igneous and metamorphic rocks are distinguished by type. Major faults are usually shown by bold lines.

Most geological maps are published by geological survey organizations. The geology of the United Kingdom is conveniently covered in two sheets at a scale of ten miles to the inch published by the British Geological Survey. Most of the country is available on larger scale maps, 1:50 000 being the commonest, which include additional information such as strike and dip, mineral veins, and so on. The equivalent organization in the USA is the United States Geological Survey, but many state geological surveys publish maps too.

A map will show you the distribution of rock types, but it will not usually tell you where these are well exposed or conveniently (and legally) accessible. This is where geological guidebooks come in handy. They usually contain sketch geological maps too, so for beginners, at least, there is often no need to invest in a large-scale geological map. Many guidebooks are written by enthusiastic geologists who know and love the area, others are more in the nature of official guides round geological trails in national parks.

Suggestions for field projects

Here are a few suggestions for geology fieldwork projects that you could tackle after reading this book. There will probably be one or two that can conveniently be attempted in your home area or on your next holiday.

- If there are quarries in your area, find out whether these occur in every local rock type. If they do not, then try to work out why some rock types are (or were) quarried and others not.

- Visit a river or stream and decide whether it is mainly eroding or depositing sediment. Does it do the same thing all year round? Could it be described as a braided or a meandering channel? Can you find any abandoned meanders (like the ox-bow lake in Figure 9.7)?

- Take your hand lens to the seaside. If it's a sandy beach, you should be able to identify quartz grains (clear, glassy) and shell fragments (milky, opaque) among the sand particles. What else can you see? Where do you think it has come from? If it's a pebbly beach, how many different rock types are represented among the pebbles? Are they all local? Do they match the rock types exposed in nearby cliffs? If not, why?

- Find an exposure with fossils in it, and try to identify them and work out how they lived (swimming, fastened to the sea floor, in burrows, etc.). Are they all of the same kind? Are most of the fossils complete or fragmentary? What can you deduce about the environment in which the fossils (i) lived and (ii) died? Are (i) and (ii) the same?

- Is there any history of mining in your area? What is or was extracted, and how was it formed? Can you find any signs of subsidence?

- What are the sources of the stone used (i) to build houses, (ii) to build churches, colleges, and municipal buildings, (iii) to decorate the facades of banks, department stores and so on, (iv) for gravestones? Is it all local, or has some been brought in from tens or hundreds of miles away? Gravestones usually bear dates; do some varieties of stone weather differently (and at different rates) from others, and if so why?

- Visit some really big hills or mountains. Why are they there? How many rock types are present? Can you find any evidence of deformation (such as folding or cleavage)? Has the landscape been eroded into its present form mainly by the action of flowing water or by ice?

The possibilities for field projects are endless, and you may think of ideas that appeal to you more than those I have suggested. Best of all, join a local geological society, which will probably offer a programme of talks and field trips.

There are many thousands of minerals known. Guides to identify even the few that are relatively common take up many pages of colour photographs and lists of properties. This appendix is not intended as a substitute for such a guide; rather it is a brief introduction that indicates some of the ways in which the very commonest minerals may be identified.

It is not straightforward to define what is meant by the term 'mineral'. A simple statement that 'a mineral is a naturally occurring crystalline substance with a well-defined chemical composition' is a good starting point. However, this is too restrictive to be an adequate definition, because the exact chemical composition of a mineral may vary, although the arrangement of atoms within its crystalline structure is fixed. On the other hand, some minerals that are regarded as distinct from each other share the same atomic arrangement, and are distinguished by differences in their chemical composition (which may radically affect their colour or other properties). Some minerals have identical chemical composition but the atoms are arranged differently, giving the crystals totally different properties (these are described as 'polymorphs'). A few substances of non-crystalline form, notably amorphous (or 'cryptocrystalline') silica such as chalcedony, flint and opal, are usually regarded as minerals too.

A thorough analysis of a mineral would require chemical analysis, study of its optical properties using a microscope, and X-ray analysis of its atomic structure. However, a common mineral occurring as a crystal bigger than a few millimetres long often shows enough clues for the moderately experienced geologist to make a fairly confident identification. There are several things to look out for. These include: lustre (does it look metallic or glassy?), colour (but impurities can make this misleading), cleavage (how many sets of cleavage planes, if any, and at what angle do they intersect?), general shape (related to cleavage in some, but not all, minerals), hardness (how easily can it be scratched?).

The hardness of minerals is judged on Mohs' scale of hardness, which runs from 1 (softest) to 10 (hardest). A harder mineral is able to scratch a softer one. Common minerals are used to define each point on the scale as show in the table below:

1	talc	can be scratched by fingernail
2	gypsum	can be scratched by fingernail
3	calcite	can be scratched by copper coin
4	fluorite	can be scratched by steel
5	apatite	can be scratched by steel
6	feldspar	scratches glass with difficulty
7	quartz	scratches glass with ease
8	topaz	scratches glass with ease
9	corundum	scratches glass with ease
10	diamond	scratches glass with ease

table A Mohs' scale of hardness

The most common rock-forming minerals, notable ore minerals, and various precious and semiprecious stones are listed in the tables below:

Name	Hardness	Distinguishing properties
Amphibole (e.g. hornblende) $(Na,K)Ca_2(Mg,Fe,Al)_5 (Al,Si)_8O_{22}(OH)_2$	5.5	Dark mineral, may be greenish, often elongated crystals, in basic igneous rocks and some metamorphic rocks, two cleavages at about 60°.

Name	Hardness	Distinguishing properties
Andalusite Al_2SiO_5	7.5	Red, often altering to silvery mica. May form cross-shaped crystals. Occurs only in metamorphic rocks. Polymorphs of the same composition are kyanite and sillimante.
Anhydrite $CaSO_4$	3–3.5	Forms pale crystals, often breaking into rectangular fragments.
Apatite $Ca_5(PO_4)_3(F,Cl,OH)$	5	The most common phosphate mineral. Crystals can take a variety of forms.
Biotite (mica) $K(Mg,Fe)_3AlSi_3O_{10}(OH)_2$	2	Virtually black, one excellent cleavage, so that it occurs as fine flakes in igneous and metamorphic rocks.
Calcite $CaCO_3$	3	Usually whitish, sometimes good crystals, three directions of cleavage (not at 90°), reacts with weak hydrochloric acid (producing bubbles of carbon dioxide).
Chlorite $(Mg,Al,Fe)_6(Si,Al)_4O_{10}(OH)_8$	2.5	Usually too fine to see, but gives green sheen to fine-grained metamorphic rocks.
Epidote $Ca_2(Al,Fe)Al_2(SiO_4)(Si_2O_7)$ $(O,OH)_2$	7	Usually green elongated crystals, in metamorphic rocks.

Name	Hardness	Distinguishing properties
Feldspar $KAlSi_3O_8$ to $NaAlSi_3O_8$ (alkali feldspar) or $NaAlSi_3O_8$ to $CaAl_2Si_2O_8$ (plagioclase feldspar)	6	Usually white or pinkish crystals, two poorly-developed cleavages roughly at 90°.
Fluorite (also known as fluorspar) CaF_2	4	Colourless to pale purple or orange, normally occurs as cubes in veins associated with ore minerals, four excellent cleavage directions.
Garnet $(Ca,Fe,Mg)_3Al_2(SiO_4)_3$	7	Red, brown or green, no cleavage, most commonly as well-formed crystals in metamorphic rock.
Gypsum $CaSO_4.2H_2O$	2	Usually whitish, can occur as good crystals, or as finely crystalline alabaster.
Illite $K_yAl_4Si_{8-y}Al_yO_{20}(OH)_4$	n/a	Illite is the most common sort of clay in mudrocks. In the formula given, the value of y is usually in the range 1 to 1.5.
Kaolinite $Al_2Si_2O_5(OH)_4$	n/a	A clay mineral with a comparatively simple chemical formula. This is the main component of china clay.

Name	Hardness	Distinguishing properties
Kyanite Al_2SiO_5	4–7	Light blue, bladed crystals that may form rosettes. Occurs only in metamorphic rocks. Polymorphs of the same composition are andalusite and sillimanite.
Montmorillonite $(Na,Ca)_{0.3}(Al,Mg,Fe)_4Si_4O_{10}$ $(OH)_2.n\,H_2O$	n/a	A clay mineral that can absorb several times its own weight of water, as implied by $n H_2O$ in its formula.
Muscovite (mica) $KAl_2(AlSi_3O_{10})(OH)_2$	2	Usually colourless, one excellent cleavage, so that it occurs as fine flakes in igneous, metamorphic and sedimentary rocks.
Olivine $(Fe,Mg)_2SiO_4$	6.5	Pale olive to very dark green, no cleavage.
Pyroxene $(Ca,Mg,Fe)_2Si_2O_6$ (augite, diopside) $(Mg,Fe)_2Si_2O_6$ (enstatite)	6	Common dark mineral in basic igneous rocks, two cleavages at about 90°.
Quartz SiO_2	7	Colourless or white, but may be tinted by impurities, no cleavage but sometimes shows nice crystal faces.
Serpentine $Mg_3Si_2O_5(OH)_4$	3–4	Has various polymorphs, including crysotile, which is fibrous and forms asbestos. A product of hydrous alteration of olivine, pyroxene and amphibole.

Name	Hardness	Distinguishing properties
Sillimanite Al_2SiO_5		Long needle-shaped brown or grey crystals. Occurs only in meta-morphic rocks. Polymorps of the same composition are andalusite and kyanite.
Talc $Mg_3Si_4O_{10}(OH)_2$	1	When occurring in bulk, talc is described as 'soapstone'. A product of hydrous alteration of olivine, pyroxene and amphibole.
Zircon $ZrSiO_4$	7.5	An accessory (rare) mineral in igneous rocks, which owing to its hardness is very robust, survives transport well and can be found in sedimentary and even some metamorphic rocks.

table B Table of common rock-forming minerals (all these have a glassy lustre).

Name	Hardness	Distinguishing properties
Cassiterite SnO_2 (tin oxide)	6	Black tetragonal more often found in crystals, water-washed sediments than within hard rock.
Chalcopyrite $CuFeS_2$ (copper-iron sulfide)	3.5	Yellow, less brassy than pyrite, tendency to tarnish.
Galena PbS (lead sulfide)	2.5	Lead-grey in colour, three good cleavage directions, may occur as cubes.
Hematite Fe_2O_3 (iron oxide)	6	Black to dark red in colour, no cleavage, may occur in kidney-like shapes.
Illmenite $FeTiO_3$ (iron-titanium oxide)	5–5.5	Black, tabular or massive crystals.
Magnetite Fe_3O_4 (iron oxide)	5.5–6	Greyish black. Common opaque mineral igneous rocks.
Molybdenite MoS_2 (molybdenum sulfide	1	Bluish-grey. Platy or massive crystals.
Pyrite FeS_2 (iron sulfide)	6	Brassy yellow in colour, often occurs as cubes, no cleavage. (This is 'fool's gold'.)
Sphalerite (zinc blende) ZnS	4	Brown, black or red, tetrahedral crystals.

table C table of common ore minerals (all these have a metallic lustre)

Name	Formula	Notes
amber	–	Arguably not a mineral. This is fossilized resin.
amethyst	SiO_2	Violet variety of quartz, coloured by impurities, notably iron.
diamond	C (pure carbon)	Formed under high pressure, this is the hardest known mineral.
emerald	$Be_3Al_2Si_6O_{18}$	A form of beryl, coloured green by minute traces of chromium.
garnet	$(Fe,Mg)_3Al_2(SiO_4)_3$	Magnesium-rich varieties are deep red and iron-rich varieties are usually violet.
jet	–	A lustrous variety of coalified fossilized wood (lignite).
lazurite	$(Na,Ca)_{4-8}(AlSiO_4)_6$ $(SO_4,S)_{1-2}$	An ornamental stone, noted for its attractive blue colour, consisting of a fine mixture of lazurite with calcite, pyroxene and other minerals.
opal	$SiO_2.nH_2O$	Amorphous, non-crystalline, hydrated silica.
ruby	Al_2O_3	A form of corundum, coloured red by traces of chromium.
sapphire	Al_2O_3	A form of corundum, coloured blue by traces of titanium and iron.
topaz	$Al_2SiO_4(F,OH)_2$	May be yellow, blue, or even red.

table D table of notable precious and semiprecious stones

appendix 2: rock names

No two rock bodies nor, for that matter, two specimens of rock collected from the same body, have exactly the same composition. Even closely similar specimens will differ imperceptibly in their chemical ingredients, or in the size or shape of the crystals or grains that make up the rock. However, a basalt and a granite are so clearly different in appearance and origin that it is both useful and sensible to have distinct names for them. This appendix lists the more common rock names and the criteria used to distinguish them.

Igneous rocks

These are rocks that formed from a molten state. Most igneous rocks are composed of an interlocking collection of crystals that grew as the melt (magma) cooled. The crystals usually have a random orientation and, generally speaking, the slower the magma cooled, the larger the crystals. Consequently, most igneous rocks can be identified as such simply on the basis of their textures. Coarse-grained igneous rocks (average crystal size greater than 2 mm across) can be identified with the unaided eye; medium-grained igneous rocks (average crystal size between 0.25 mm and 2 mm) usually need a hand lens or magnifying glass; and fine-grained igneous rocks (average crystal size less than 0.25 mm) require a microscope to be certain of their nature and composition.

There are several notable exceptions to these general remarks:

- flow or settling during crystallization can cause crystals to become oriented in layers

- some volcanic igneous rocks may have cooled so rapidly that no crystals had time to grow, with the result that the rock has the appearance, and sub-microscopic structure, of glass
- volcanic rocks that were erupted explosively (pyroclastic rocks) are made of either glassy or crystalline fragments
- the very coarse grain size of rocks described as pegmatites (see Chapter 06) bears no relation to the rate of cooling.

Whatever their texture, igneous rocks are formally classified on the basis of their chemical composition, notably by the proportion of silica (SiO_2) in their total chemical make up. Chemical analyses cannot usually be made in the field, but the overall chemistry of a rock is reflected in the kinds of minerals it contains so, fortunately, it is possible to identify rock types in this way. The more silica a rock has, the more acidic it is said to be, and igneous rocks are divided into acidic (otherwise known as 'felsic'), intermediate, basic (or 'mafic') and ultrabasic (or 'ultramafic') types.

Acid (felsic) igneous rocks

These contain greater than 66 per cent silica (SiO_2), which is not to say that they contain greater than 66 per cent quartz (the mineral with SiO_2 as its formula), because all silicate minerals contain some SiO_2 in their formulae. The most abundant minerals in an acidic igneous rock are alkali feldspar, quartz, and plagioclase feldspar, with lesser amounts of muscovite, biotite and sometimes amphibole. If the rock is coarse grained it is called granite, if it is medium grained it is called microgranite, and if it is fine grained it is called rhyolite. The term 'granitic' is used to describe acidic igneous rocks in general; for example, a rhyolite is a fine-grained granitic igneous rock. A pyroclastic rock of this composition would be referred to as a 'rhyolitic tuff'.

Intermediate igneous rocks

These contain between 52 per cent and 66 per cent silica. The most abundant mineral is plagioclase feldspar, with lesser amounts of biotite and amphibole, and sometimes alkali feldspar, quartz or pyroxene. A coarse-grained intermediate igneous rock is called diorite, its medium-grained relative is called microdiorite, and the fine-grained equivalent is called andesite. The term 'andesitic' is used to describe these rocks in general, so that a diorite can be said to be a coarse-grained andesitic rock.

Basic (mafic) igneous rocks

These contain between 45 per cent and 52 per cent silica (but the mineral quartz is entirely absent). The most abundant minerals are plagioclase feldspar and pyroxene, and sometimes olivine. A coarse-grained basic igneous rock is called a gabbro, its medium-grained equivalent is called a dolerite (British usage) or diabase (American usage), and the fine-grained variety is called basalt. The term 'basaltic' is used to describe these rocks in general.

Ultrabasic (ultramafic) igneous rocks

These contain less than 45 per cent silica (usually not reaching less than about 38 per cent). The most abundant mineral is olivine, with pyroxene and a lesser amount of plagioclase feldspar. Only the coarse-grained variety is at all common, and is named peridotite.

Other distinctions

The names highlighted above are just the most common names, and there is a bewildering variety of names still in use. Some of these are inherited from the days when each rock type was named after the place it was found (for example, a certain variety of basalt is still referred to as 'hawaiite', after Hawaii). Other names recognize that silica content is not the only significant variable in rock composition: another is the relative abundance of the 'alkali' metals sodium and potassium (which tend to go into alkali feldspar) compared with calcium (which goes into plagioclase feldspar). Rock names such as syenite, monzonite and dacite reflect a decreasing abundance of alkalis in alkali-rich rocks that are broadly of granitic to andesitic composition.

Metamorphic rocks

These are rocks that form when heat or pressure, or both, cause a pre-existing rock to recrystallize. Very often, the effect of pressure is to cause crystals to grow in, or to re-orient themselves into, layers. When this happens the metamorphic rock has a layered texture, which is described as 'foliation'. In fine-grained metamorphic rocks the crystals cannot be seen by the unaided eye, but the foliation is manifested by the tendency of the rock to fracture along flat or wavy 'cleavage' surfaces parallel to the foliation.

Metamorphic rocks are most easily classified simply on textural grounds, using the same fine-, medium- and coarse-grain size division as in igneous rocks, as shown in the following table.

Textural type	Nature of foliation	Grain size
slate	Very closely spaced, almost flat	fine
phyllite	Wavey or crenulated, sheen caused by minute flakes of mica or chlorite on foliation surfaces	fine
schist	Undulating planes, characteristically with abundant mica	medium to coarse
gneiss	More widely spaced than in schist, usually alternating layers of dark and light minerals	coarse
migmatite	Poor foliation, shows signs of having begun to melt	coarse

table X Classification of metamorphic rocks by texture.

It is often convenient to qualify the simple terms listed above in recognition of prominent minerals within the rock. For example, a schist with abundant mica and distinct crystals of garnet may be described as a garnet mica schist.

Not all metamorphic rocks have a foliated texture. For example, a rock metamorphosed by the heat of an igneous intrusion (contact metamorphism) usually develops a spotty texture and is described as a hornfels. Furthermore, rocks that, prior to metamorphism, consisted of just one abundant mineral are unlikely to develop a new mineral assemblage during metamorphism. The original texture may be destroyed by adjustment of crystal boundaries, but there will not be a strong oriented fabric. A quartz-rich sandstone that has been metamorphosed consists of interlocking quartz crystals and is described as quartzite, and a limestone that has been metamorphosed consists of interlocking calcite crystals and is described as marble.

Chapter 07 introduced the concept of metamorphic facies, corresponding to the temperature and pressure under which metamorphism took place. The facies cannot usually be determined by simple inspection; however, when the facies is known then it is normal to write of (say) an amphibolite facies gneiss or a zeolite facies slate.

Sedimentary rocks

These form by deposition of detrital grains or by precipitation from solution in water. Detrital (or 'fragmental') sediments are most simply distinguished from igneous and metamorphic rocks on textural grounds, because the grains of which they are composed are non-interlocking fragments rather than interlocking crystals. Sediments formed by precipitation do have interlocking cystalline textures, but are made of quite different minerals from igneous rocks, notably calcite (in limestone) or evaporite minerals such as gypsum.

Fragmental sediments are classified according to grain size in the scheme shown in the following table, which shows names for the rock type and the different sizes of clasts (note: 'breccia' comes from the Italian and is pronounced 'brechia').

Grain size	Term for clasts	Rock type
>256 mm	boulders	**conglomerate** (rounded clasts) or **breccia** (angular clasts)
64–256 mm	cobbles	**conglomerate** (rounded clasts) or **breccia** (angular clasts)
2–64 mm	gravel (pebbles 4–64 mm)	**conglomerate** (rounded clasts) or **breccia** (angular clasts)
0.0625–2 mm	sand	**sandstone**
0.002–0.0625 mm	silt	**siltstone**, or **mudrock** (referred to as **shale** if it splits easily into layers)
<0.002 mm	clay	**claystone**, or **mudrock** (referred to as **shale** if it splits easily into layers)

Table Y Classification of sedimentary rocks by grain size.

Rocks of these compositions may sometimes also be described according to their supposed mode of origin (which is a risky business, because the geologist may get this wrong!). For example, a breccia or conglomerate with a muddy matrix that is thought to have been deposited by a glacier would be called a tillite.

Sandstones are further classified according to the mineralogy of their grains. A sandstone that is over 90 per cent quartz grains is described as a quartz sandstone, whereas one containing a significant amount of feldspar, clay minerals or calcite is called an arkose, muddy sandstone or calcarous sandstone, respectively. One that has a significant proportion of fragments composed of fine-grained rocks is described as a greywacke.

Just as a hardened rock made of sand is referred to as a sandstone, so a rock made from silt is referred to as a siltstone and one made from clay is a claystone. See table Y for these and other terms.

A rock that consists of over 90 per cent calcite is referred to as a limestone. A limestone with an admixture of quartz grains or clay minerals would be called a sandy limestone or a muddy limestone. Pure limestones are further distinguished according to whether the calcite occurs as fragments of shells (skeletal limestones) or as ooids (oolitic limestone), and whether the grains are cemented together by crystalline calcite (sparite) or held together by fine-grained calcite mud (micrite). If the calcite has been replaced by the mineral dolomite, the rock should be called a dolostone, but is more commonly referred to as dolomite.

Another common sedimentary rock name is chert, which describes a rock formed of microscopic silica crystals (which may be a chemical precipitate, as in the case of flint, or shells of microscopic organisms such as radiolarians).

Geological time is divided according to the scheme shown in the diagram overleaf. Note that this is not drawn to a uniform scale. The numbers refer to the age in millions of years of the boundary between each division, as determined radiometrically (i.e. by measuring the products of radioactive decay). These dates are still subject to revision.

The largest unit of geological time is an eon. The scheme shown here names four eons. The oldest known rocks date back to the beginning of the Archean, and the Hadean is that time before for which we have only indirect evidence. Some authorities do not regard the Hadean as a separate eon, and include this time within the Archean.

The Phanerozoic Eon is divided into three eras – Paleozoic, Mesozoic and Cenozoic (which mean old-life, middle-life and modern-life respectively). In many kinds of sedimentary rocks of Phanerozoic age, fossils are sufficiently abundant and show sufficiently clear evolutionary change to enable geological time to be split into a succession of progressively smaller divisions. Each era is divided into a number of periods. Each period is divided into a number of epochs (shown here only for the Paleogene and Neogene Periods), and further subdivided into ages (not shown here).

The Holocene Epoch covers the past 10 000 years, and is sometimes called the Recent. The time span covered by the Cenozoic was formerly split into two eras – the Tertiary Era (from the start of the Paleocene to the end of the Pliocene) and the Quaternary Era (Pleistocene and Holocene), and both these terms are still in widespread use. The boundary between the Cretaceous and the Tertiary, which is marked by the most recent major episode of mass extinction, is often referred to as the K/T boundary (K being the usual abbreviation for Cretaceous).

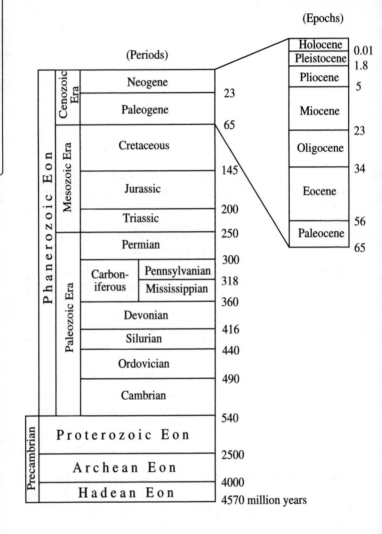

glossary

This glossary contains those terms that I think the newcomer to geology will have most trouble with. I have indicated the chapter(s) in which each term makes an important appearance. Names of several minerals and rock types not listed here may be found in Appendices 1 and 2. There is a stratigraphic timescale in Appendix 3.

acidic Descriptive term for igneous rocks containing more than 66 per cent silica (SiO_2). Alternatively referred to as 'felsic' (Ch 05).

andesite A fine-grained igneous rock type richer in silica than basalt, commonly erupted at volcanoes above subduction zones (Ch 05).

anticline An arch-like fold, in which beds on either side dip away from each other (Ch 10).

ash Fine, fragmentary rock, produced by a pyroclastic (explosive) eruption (Ch 05).

asthenosphere The weak zone within the mantle (but below its top), which is overlain by the lithosphere (Ch 02).

banded iron formation (BIF) A sedimentary deposit consisting of alternate layers of iron oxide and silica, of the kind constituting most of the global iron reserves. Most are between 3800 and 1800 million years old (Ch 11).

basalt An igneous rock type with the same composition as oceanic crust. Strictly speaking, a basalt must be fine grained; coarser-grained rocks of the same composition are described as basaltic or basic (Ch 02, 04).

basic Descriptive term for igneous rocks containing more than 45–52 per cent silica (SiO_2). Alternatively referred to as 'mafic' (Ch 04, 06).

bed The commonest and simplest term for a layer of sedimentary rock. Also used to refer to the floor of the sea or the surface over which wind or water is flowing (Ch 09).

bedform A ripple, dune or other similar feature produced by flowing water or wind (Ch 09).

bed-load Particles dragged, rolled or bounced along during transport by wind or water (Ch 08).

biostratigraphy Determining the (relative) ages of sedimentary rocks by means on their fossil content (Ch 12).

breccia A rock composed of coarse angular fragments. Can be a sediment or a product of intense local fracturing during faulting (Ch 10).

brittle Deformation of rock by fracturing. A characteristic of rapid deformation at relatively low temperatures and pressures (Ch 10). Contrast **ductile**.

caldera A volcanic crater more than 1 km across, formed by subsidence of its floor during an eruption (Ch 05).

carbon cycle A term describing the complex natural cycle in which carbon is transferred between the atmosphere (as carbon dioxide), the oceans (in solution or in the shells of living organisms) and the Earth's interior (as carbonate rocks, as hydrocarbons or as carbon dioxide emitted by volcanoes) (Ch 01).

cement The precipitated mineral holding grains together in a sedimentary rock, most commonly calcite or silica (Ch 09).

chilled margin The very fine-grained/glassy edge of a shallow igneous intrusion, which cooled rapidly against the cold rock into which it was injected (Ch 06).

clast A particle or lump (of any size) that is transported and deposited in a sediment (Ch 09).

clastic rock A sedimentary rock composed of clasts (Ch 09).

cleavage Regular planes of weakness in a crystal, which are a consequence of its atomic structure (Ch 06, 08).

The same term describes closely spaced planes of weakness in a deformed or metamorphosed rock, which are caused by alignments of platy minerals (Ch 07, 10).

conservative plate boundary A site where two tectonic plates are sliding past each other. In the oceans, the fault across which the motion occurs is called a transform fault (Ch 04).

constructive plate boundary A site where two tectonic plates are moving apart, and each is being added to by the formation of new oceanic lithosphere (Ch 04).

continental drift The relative movement of continents, which is a consequence of plate tectonics (Ch 04).

core The Earth's high-density iron-rich interior (3500 km in radius). It consists of a liquid outer core and a solid inner core (Ch 02).

craton A stable area of continental crust, also known as a shield, that has remained undeformed since the Archean Eon (Ch 13).

cross-bedding Sedimentary bedforms seen in cross-section, where the dip of the incrementally added layers was originally steeper than the dip of the bed as a whole (Ch 09).

crust The compositionally distinct, slightly lower density rocky layer overlying the Earth's mantle (Ch 02).

crystal A solid substance formed by a repeating, three-dimensional pattern of atoms of various elements. The symmetry of this pattern is reflected in the characteristic shape of the crystal. Most minerals occur in crystalline forms (Ch 01, 06).

decompression melting Partial melting stimulated in the asthenosphere when mantle rises upwards (in a plume or below a constructive plate boundary). Melting is caused by the drop in pressure rather than the addition of heat (Ch 04).

destructive plate boundary A site where one tectonic plate is destroyed by subduction below another (Ch 04).

diagenesis The chemical and physical processes that turn an unconsolidated sediment into a hard sedimentary rock (Ch 09).

dip The angle between the horizontal plane and a geological surface, such as a bedding plane (Ch 15).

ductile Deformation of rock by bending or squeezing, but without fracturing. A characteristic of slow deformation at relatively high temperatures and pressures (Ch 10). Contrast **brittle**.

dyke A curtain of igneous rock, intruded vertically (Ch 06).

earthquake Shaking of the ground, caused by sudden movement within the Earth (Ch 02, 03).

epicentre The point on the Earth's surface directly above an earthquake's focus (Ch 03).

evaporite A mineral or deposit of minerals formed by evaporation of seawater (Ch 09).

evolution A theory, famously propounded by Charles Darwin in the nineteenth century and subsequently backed up by a great weight of genetic and fossil evidence, that pressure to survive in a competitive environment increases the chances of individuals with successful adaptations surviving to breed, and hence to pass on their adaptations to their progeny. Life has diversified from the first primitive microbes to its present state by the slow accumulation of random mutations that gave competitive advantage in particular environments (Ch 12).

exposure A place where bedrock is actually visible at the surface. Sometimes casually referred to as an **outcrop**, though, strictly, that term has a different meaning (Ch 15).

facies The temperature and pressure conditions under which metamorphic rocks were metamorphosed (Ch 07).

fault a fracture in the lithosphere across which motion takes place (Ch 03).

fieldwork studying rocks and geological processes 'in the field', meaning 'where they occur' (Ch 15).

flood basalt Extensive, thick pile of flat-lying basalt lava flows (typically more than a million cubic kilometres in volume), thought to be generated above the site of a particularly vigorous mantle plume (Ch 05).

flood plain Flat, low-lying ground beside a river, over which it floods after very heavy rainfall. A river's course tends to migrate to and fro across its flood plain over tens of thousands of years (Ch 09).

focus The break-point where earthquake motion begins, and where most of the energy is released (Ch 03).

fold A structure formed when layers of rock become bent (Ch 10).

foliation An alignment of minerals in layers, caused by regional metamorphism (Ch 07).

footwall The volume of rock below an inclined fault surface, and which is not deformed when the overlying rock is displaced by fault motion (Ch 10).

fossil The remains of a past organism. A fossil is either a body fossil, being a fossilized body part, or a trace fossil such as a footprint or burrow (Ch 12).

fossil fuels Coal, oil and natural gas, which are being used up at rates much faster than the rates at which they are being created by geological processes (Ch 11).

fractional crystallization The process whereby the first crystals to form in a magma as it cools have a composition different from that of the magma as a whole (Ch 06).

geothermal gradient The rate at which temperature increases with depth inside the Earth (Ch 02).

glaciations Cold periods (tens or hundreds of thousands of years) during an ice age when ice sheets are extensive, and punctuated by interglacial periods (Ch 08). The term is sometimes wrongly equated with 'ice age'.

glass Volcanic glass is magma that cooled so quickly that crystals had no time to form (Ch 06).

gneiss High-grade metamorphic rock characterized by alternating dark and pale layers (Ch 07).

granite A coarse-grained intrusive igneous rock, richer in silica than andesite. Rocks of this composition are described as granitic (Ch 05, 06).

greenhouse effect The warming of a planet's atmosphere by means of absorption of infrared radiation by certain atmospheric gases, notably carbon dioxide (Ch 09, 11, 13).

hangingwall The volume of rock above an inclined fault surface, and which may be deformed when the rock is displaced by fault motion (Ch 10).

hornfels A splintery, spotty rock produced by contact metamorphism (Ch 07).

hot spot A place where a plume rising from deep within the mantle hits the base of the lithosphere, giving rise to large volumes of basaltic magma (Ch 05).

hydration melting Partial melting simulated by the addition of water, notably where water escapes from a subducting plate into the mantle wedge above (Ch 05).

hydrocarbons Oil and natural gas, consisting of chains of carbon atoms bonded with hydrogen (Ch 11).

hydrological cycle A term describing the complex natural cycle in which water is transferred between the atmosphere, the oceans and the Earth's interior (Ch 01).

hydrolysis Chemical breakdown of a mineral to clay particles by the action of (usually slightly acidic) water during weathering (Ch 8).

ice age A period (of the order of ten million years long) when the climate alternates between colder and warmer, so there is a succession of glaciations (when polar ice sheets and mountain glaciers are particularly extensive) and interglacials (when, as at present, the ice has retreated) (Ch 08, 13).

igneous Referring to a rock or mineral formed by solidification from a molten state (Ch 01).

ignimbrite An extensive pyroclastic deposit formed from shards of acidic volcanic ash and pumice fragments, usually welded together. Typically emplaced from a pyroclastic flow as a result of the collapse of an eruption column during the eruption of a supervolcano, or a smaller eruption of similar type.

interglacials Interludes of relatively warm climate between successive glaciations of an ice age (Ch 08).

intermediate Descriptive term for igneous rocks containing 52–66 per cent silica (SiO_2), such as an andesite (Ch 05).

intrusive Referring to an igneous rock emplaced below the surface (Ch 06).

island arc A usually arcuate line of islands built by volcanism above a subduction zone (Ch 04).

isostasy Really just a way of saying 'buoyant equilibrium'. It is normally found that mountain regions have deep roots. In contrast the Moho (the base of the crust) is further from the centre of the Earth below low-lying regions. Seen from a geological perspective, the crust thus 'floats' in equilibrium upon the mantle, although on human timescales the mantle behaves as a solid rather than a liquid (Ch 02).

joint A smooth fracture through a body of rock, caused by release of pressure or cooling. In contrast to a **fault**, there is no movement between opposite sides of a joint (Ch 08).

K-T boundary The boundary between the Cretaceous (K) and the younger sediments of Cenozoic or Tertiary (T) Age. It coincides with an important mass extinction event, which was probably caused by the impact of an asteroid or comet (Ch 12).

lava Molten rock at the surface. The term is also used to describe rock formed in this way after it has solidified (Ch 05).

lithosphere The rigid outer shell of the Earth, consisting of the crust and the uppermost mantle (Ch 02).

limestone Rock made mostly of calcium carbonate (Ch 07, 08, 09).

magma Molten rock; usually used to refer to molten rock at depth, whereas molten rock at the surface is called lava. Magma often includes dispersed crystals, so the term 'melt' is sometimes used when referring specifically to the liquid component (Ch 04).

mantle All the silicate (rocky) material surrounding the core of the Earth, with the exception of the thin silicate crust (Ch 02).

mass extinction A relatively brief interlude (tens or hundreds of thousands of years) during which a high proportion of species of many kinds become extinct. The K-T boundary mass extinction was probably a result of an asteroid impact (Ch 12).

metamorphic Referring to a rock or mineral formed by recrystallization, without melting but in response to heat and/or pressure, of a pre-existing rock or mineral. This process is called metamorphism, and can usually be described as either thermal or regional (Ch 01, 07).

metamorphism Recrystallization of the fabric of a rock without melting, under conditions of high temperature and/or pressure. Thermal or contact metamorphism is a result of proximity to an igneous intrusion. Regional metamorphism is a result of burial to great depth (Ch 07).

mineral A naturally occurring crystalline substance with a well-defined chemical composition. Most rock types consist of several different minerals (Ch 06). See Appendix 1 for a fuller discussion.

Moho The boundary between the crust and the mantle (Ch 02).

non-renewable resources Fossil fuels, ores and bulk materials, which are being used up at rates much faster than the rates at which they are being created by geological processes (Ch 11).

normal fault Steep fault with vertical displacement, where the rock on the down-dip side of the fault has moved downwards (Ch 10).

ophiolite A slice of oceanic crust and upper mantle that has been thrust over the edge of a continent (Ch 04).

ore A metal-bearing mineral from which the metal can be extracted at a profit (Ch 11).

orogeny A mountain-building episode resulting from a collision between two continents, or between a continent and an island arc (Ch 13).

outcrop On a geological map, the area where a particular unit of rock would be visible if all superficial cover were removed (Ch 15). Contrast **exposure**.

partial melting What happens when any rock composed of a variety of minerals begins to melt. The first melt to form has a different chemical composition (richer in silica) from the solid (Ch 04).

peridotite A silicate rock type poor in silica, having the same composition as the Earth's mantle (Ch 02, 04, 06).

permeability The ease with which fluid or gas can flow through a porous rock, which is dependent on how well-connected the pore-spaces are (Ch 11).Contrast **porosity**.

phenocryst A larger crystal surrounded by smaller crystals in an igneous rock (Ch 06).

pillow lava The characteristic morphology shown by basaltic lava when erupted under water (Ch 05).

placer deposit An ore concentrated in a sediment, because the ore forming the densest grains becomes concentrated by sedimentary processes (Ch 11).

plate tectonics A description of how the rigid plates into which the Earth's lithosphere is divided move about (Ch 04).

polymorphs Minerals that have identical chemical formulae but different crystalline structures. Examples are kyanite, sillimanite and andalusite (Al_2SiO_5) and calcite and aragonite ($CaCO_3$) (Ch 07).

porosity The proportion of a rock body that is occupied by liquid- or gas-filled spaces (Ch 09, 11). Contrast **permeability**.

prograding Description of a shoreline that is building out seawards (Ch 09).

pyroclastic Term used to describe an eruption (or the resulting rock) in which fragments are produced by explosive volcanic activity (Ch 05).

pyroclastic flow A denser than air mixture of ash and larger bouncing rock fragments mixed with hot air that flows turbulently downhill (Ch 05).

radiogenic heating Heating of a planet's interior through the energy released by decay of the nuclei of unstable, radioactive atoms (Ch 02).

radiometric dating A technique in which the relative proportions of products of radioactive decay and their parent isotopes are used to determine the age of a rock or mineral (Ch 02).

regional metamorphism Metamorphism as a result of pressure and heat at depth within the crust, which can be seen in formerly deep rock that has been exposed at the surface as a result of uplift and erosion (Ch 07).

reserves Those resources that can be extracted profitably and legally under existing conditions (Ch 11). Contrast **resources**.

reservoir rock A permeable body of rock into which hydrocarbons or water have migrated, and from which they can be extracted through a well (Ch 11). Contrast **source rock**.

resources The estimated total amount of a commodity (fossil fuel, ore or raw material) available in the world (Ch 11). Contrast **reserves**.

reverse fault Steep fault with vertical displacement, where the rock on the down-dip side of the fault has moved upwards (Ch 10).

rock cycle A complex network of processes including the erosion, transport, deposition, burial, heating, deformation, melting, cooling and exhumation of rock or rock fragments (Ch 01).

schist Metamorphic rock characterized by wavy layers of minerals, whose crystals can be distinguished by eye (Ch 07).

sedimentary Referring to a rock formed from a deposit of detrital clasts, or by precipitation from solution in water. Sedimentary minerals are those that form in, or are found in, sedimentary rocks (Ch 01).

seismic waves Vibrations, from earthquakes or artificial explosions, that travel through the Earth. P-waves are compression-dilation waves, and S-waves are shearing waves (Ch 02).

seismometer Device for recording the vibrations caused by earthquakes or movement of magma (Ch 02, 03, 05).

shale A mudrock or siltstone that breaks naturally into finely-spaced layers parallel to its bedding (Ch 09, 11, Appendix 2).

shield Another world for **craton** (Ch 13).

shield volcano A gently-sloping basaltic volcano, so named because of its cross-sectional profile (Ch 05).

silica Any compound with the formula SiO_2. The term is also used to refer to the percentage of a rock (by weight) that can be expressed as SiO_2 irrespective of whether this occurs as pure silica or combined with other elements in silicate minerals (Ch 02).

silicate Either a rock type or a mineral rich in silicon and oxygen. The Earth's mantle and most of its crust are formed of silicates, as are virtually all igneous rocks (Ch 02, 06).

sill A horizontal sheet-like igneous intrusion (Ch 06).

slate Low-grade, fine-grained, metamorphic rock, prone to fracturing on planes parallel to its imposed foliation (Ch 07).

source rock A body of rock within which hydrocarbons form (Ch 11). Contrast **reservoir rock.**

strata General term for layers of rock, irrespective of whether they are sedimentary. The singular is 'stratum' (Ch 10).

strike The direction, conventionally measured clockwise from north, of a horizontal line drawn on a geological surface, such as a bedding plane (Ch 15).

subduction Term describing one tectonic plate descending at an angle below another, which happens at a destructive plate boundary (Ch 04).

suspension-load Particles suspended for long periods without touching the bottom during transport by wind or water (Ch 8).

syncline A U-shaped fold, in which beds on either side dip towards each other (Ch 10).

terrane A fault-bounded region of the crust that has a different geological history to that of the adjacent crust, because they were formerly not neighbours. (The alternative spelling, 'terrain', refers to landscape in general.)

thermal metamorphism Metamorphism of pre-existing rock near the margin of an igneous intrusion, and driven by heat leaking out of it. Also known as 'contact metamorphism' (Ch 07).

thrust A low-angle fault which has pushed older rocks over younger rocks (Ch 10).

transcurrent fault A steep fault across which relative displacement is sideways. Alternatively known as a strike-slip fault (Ch 10).

transform fault A fault in oceanic lithosphere offsetting two lengths of a constructive plate boundary, which has sideways slip across it (Ch 04).

tsunami A series of water waves triggered by an earthquake, landslide or volcanic eruption. The waves' height is slight while they travel through deep water, but becomes much higher and extremely dangerous as they approach the shore (Ch 03, 05).

ultrabasic Term describing rock that has even less silica than a basalt (<45 per cent), such as a peridotite. Alternatively referred to as ultramafic (Ch 06).

unconformity A boundary between underlying older rocks and an overlying set of younger strata, usually manifested by a difference in dip and representing a considerable time gap (Ch 10).

vein A fracture that has become filled by minerals precipitated from solution, commonly quartz or calcite, and sometimes including ore minerals (Ch 07, 11).

Coarse veins described as pegmatite may crystallize from the last fraction of melt remaining as an igneous intrusion cools down (Ch 06).

volcano A landscape feature where magma is erupted (Ch 02, 05).

weathering Slow decay of rock and its constituent minerals upon exposure at the Earth's surface, by either physical or chemical attack (Ch 07, 08).

taking it further

If you have enjoyed this book and would like to discover more about geological natural hazards you will find much more about them in *Teach Yourself Volcanoes, Earthquakes and Tsunamis*, by the same author.

Websites

General geology

http://www.si.edu/science_and_technology/ A site at the Smithsonian Museum of Natural History (Washington, DC). Lots of good things to explore if you click on 'Geology'.

http://www.usgs.gov/ The home page of the United States Geological Survey; complicated but you can find information on most geological topics here.

http://www.bgs.ac.uk/ The home page of the British Geological Survey.

http://www.gsi.ie/ The home page of the Geological Survey of Ireland.

http://www.agso.gov.au/ The home page of Geoscience Australia.

http://www.gns.cri.nz/ The home page of New Zealand's Institute of Geological and Nuclear Sciences.

http://www.geolsoc.org.uk/ The home page of the Geological Society of London, the organization to which most professional geologists in the UK belong.

http://www.geologists.org.uk/ The home page of the Geologists Association, representing amateur geologists in the UK.

http://www.ougs.org.uk/ The home page of the Open University Geological Society, a Europe-wide self-help group for serious or casual students of geology. News of local meetings and field trips.

http://geology.about.com/ A free commercial site (with advertising) offering up-to-date news and web links across the full range of geology, including pictures of rocks, minerals and fossils.

Minerals

http://webmineral.com/ A database of minerals, including photographs and analyses.

http://www.minersoc.org/ The website of the Mineralogical Society (UK). Includes a picture gallery of spectacular mineral specimens, and links to other useful sites.

http://www.minsocam.org/msa/collectors_corner/index.htm The 'collector's corner' at the website of the Mineralogical Society of America. Contains pictures of minerals and guidance on rock and mineral collecting.

Fossils

http://www.nhm.ac.uk/ The website of the Natural History Museum (London, UK). Provides access to a wide variety of information about fossils, including some interactive 'virtual reality' three-dimensional models.

Earthquakes

http://earthquake.usgs.gov/ The website of the United States Geological Survey Earthquake Hazards Program. Provides global and regional maps of earthquake locations, news of the largest and most significant earthquakes, and links to many other useful earthquake websites.

http://earthquake.usgs.gov/4kids/ Earthquake information specially for kids.

http://www.earthquakes.bgs.ac.uk/ The earthquake website of the British Geological Survey.

Volcanoes

http://www.volcano.si.edu/ The website of the Global Volcanism Program. The most comprehensive site for recent news and archives of information about volcanic activity around the world.

http://volcanoes.usgs.gov/ The website of the United States Geological Survey Volcano Hazards Program. Details on US and Russian volcanoes, and weekly summaries of current global activity.

http://volcano.und.edu/ 'VolcanoWorld' another good place to start browsing for volcanoes, and specially suitable for kids.

Natural resources

Annual statistics on global production of metals and other commodities extracted by mining can be found at http://minerals.usgs.gov/minerals/. Similar information about coal, oil and gas production can be found at http://energy.usgs.gov/

Earth history

Maps and animations of plate tectonic movements and the changing pattern of global climate can be found at http://www.scotese.com/